Energy Storage

A Nontechnical Guide

Energy Storage

A Nontechnical Guide

By Richard Baxter

Copyright© 2006 by
PennWell Corporation
1421 South Sheridan Road
Tulsa, Oklahoma 74112-6600 USA

800.752.9764
+1.918.831.9421
sales@pennwell.com
www.pennwellbooks.com
www.pennwell.com

Managing Editor: Steve Hill
Production/Operations Manager: Traci Huntsman
Production Manager: Robin Remaley
Assistant Editor: Amethyst Hensley
Book Design: Wes Rowell

Library of Congress Cataloging-in-Publication Data Available on Request

Baxter, Richard
  Energy Storage: A Nontechnical Guide
     p. cm.
  ISBN 1-59370-027-X

Printed in the United States of America

1 2 3 4 5   10 09 08 07 06 05

Contents

x

List of Figures

1. Storage and the Electric Power Industry

2. Storage in Other Energy Markets

3. Electricity Storage Technologies

4. Applications

5. Renewable Energy and Storage

6. Our New Energy Future

Acronyms and Abbreviations

AC	Alternating current
ATC	Available transfer capability
Btu	British thermal unit
CAES	Compressed air energy storage
CAO	Control area operator
C&I	Commercial and industrial
DC	Direct current
DG	Distributed generation
DOD	Depth-of-discharge
DOE	Department of Energy
DR	Demand response
DSM	Demand side management
EEI	Edison Electric Institute
EIA	Energy Information Agency
EPA	Environmental Protection Agency
EPAct	Energy Policy Act
EPC	Engineering, procurement, and construction
EPRI	Electric Power Research Institute
ERCOT	Electric Reliability Council of Texas
ESA	Energy Storage Association
ESC	Energy Storage Council
ESS	Energy Storage Systems
FACTS	Flexible AC current transmission system
FERC	Federal Energy Regulatory Commission
GW	Gigawatt
GWh	Gigawatt hour
HTS	High-temperature superconductivity
HVDC	High voltage direct current
IEA	International Energy Agency
ISO	Independent system operator
kVA	Kilo-volt-ampere
kW	Kilowatt
kWh	Kilowatt-hour

LDC	Local distribution company
LTS	Low-temperature superconductivity
MISO	Midwest independent system operator
MJ	Mega joule (1MJ=0.28 kWh)
MRO	Maintenance, repair, and operation
MVA	Mega-volt-ampere
MVAR	Megavars
MW	Megawatt
MWh	Megawatt hour
NAS	Sodium sulfur
NEP	National Energy Policy
NERC	North American Electric Reliability Council
NETL	National Energy Technology Laboratory
NGA	Natural Gas Act
NiCd	Nickel cadmium
NREL	National Renewable Energy Laboratory
NOx	Nitrous oxides
NYMEX	New York Mercantile Exchange
O&M	Operation and maintenance
OMB	Office of Management and Budget
ORNL	Oak Ridge National Laboratory
PCS	Power conversion system
PF	Power factor
PHS	Pumped-hydro (electric) storage
PJM	Pennsylvania–New Jersey–Maryland Interconnection
PUC	Public Utility Commission
PURPA	Public Utility Regulatory Policy Act
R&D	Research & development
RTO	Regional Transmission Organization
SMES	Superconducting magnetic energy storage
SNL	Sandia National Laboratory
SO_2	Sulfur dioxide
T&D	Transmission & distribution
Tcf	Trillion cubic feet
TES	Thermal energy storage
UPS	Uninterruptible power supply
US	United States
VAR	Volt-ampere reactive
VRLA	Valve-regulated lead acid
WTG	Wind turbine generator

Foreword

Electricity is the most useful and flexible of all energy sources. To provide this capability, the power industry in modern industrialized societies developed power stations of various sizes and capabilities to provide a continuous, reliable, and affordable supply of electricity as the demand varied on daily, weekly, and seasonal cycles. Lately however, this centrally organized and controlled market design has become unstable. This has caused investment for new large power stations to become riskier as repaying their development costs can no longer be guaranteed through assured power sales in a highly regulated market.

Solutions to this challenge follow one of two competing strategies. The first is simply to continue extending the power transmission grid in order to open up additional markets for these new generation facilities. The second is to focus on a distributed supply strategy reliant upon smaller and distributed power generation and energy storage resources to provide a more stable and secure electricity supply. Each of these strategies implies a different direction for the future of the power industry. The first strategy represents a continued centralization of power production to offset increasing transmission infrastructure costs, whereas the second strategy represents a focus on local production and management of electricity to avoid excessive infrastructure build-out and grid management costs. This second option is the most promising one as it enables significant progression toward improvements of energy efficiency, leads to enhanced energy security, and—above all—promotes wind and solar power for the electricity supply.

Richard Baxter shows in his book how diverse the possibilities of electricity storage really are. These technologies have been widely overlooked by the power industry for many years as the industry's focus has been fixated on large-scale supply strategies. Through this fundamental book for the energy industry of tomorrow, Richard Baxter has broadened the industry's horizon by showing how it will be revolutionized by energy storage technologies—enabling greater use of renewable energy, and promoting a more flexible, efficient, and stable self-correcting energy infrastructure.

Dr. Hermann Scheer
General Chairman World Council for Renewable Energy
President EURSOLAR
Recipient of the Alternative Nobel Prize
Member of the German Bundestag

1 — STORAGE AND THE ELECTRIC POWER INDUSTRY

The electric power industry has some immense challenges before it—that, we can all agree, is glaringly obvious; the good news is that energy storage technologies offer real solutions to some of the most pressing of these issues. Many of the worst problems stem from issues built into the system through the market structure. One of most vexing is that the current power system is built around a central tenet: Electricity must be produced when it is needed and used once it is produced. This rule necessitates rigid procedures for operating the system—raising inefficiency, lowering reliability, and reducing security. Although radical solutions from pundits abound, most industry veterans understand the sheer scale and interconnectivity of the system mean that change here comes most readily through evolutionary and not revolutionary means. Because energy storage technologies are usually enabling technologies and not disruptive ones, their expanded use will enhance the value of existing assets by providing more flexibility and options—supporting the inherent infrastructure-centric nature of the market.

Although the industry may prefer slow, evolutionary change, it should also be understood that change is endemic in the power industry. In the short-term operation of the electric power market, what happens on a daily basis can take weeks or months in the natural gas market. In the long term, evolving regulatory, economic, and technological forces have affected the market since its inception, and will continue to do so in the future; while some of this change is intended, much is either unexpected or even unwanted. First, it exposes weakness in the market; episodes like the 2003 blackout in Canada and the northeastern United States show that there is a real need for the existing infrastructure to be coordinated in a far more effective manner. Second, change creates a need for new market tools—existing assets are built to perform in certain market conditions, and when these change, some of these assets are simply not readily adaptable. Finally, change also produces opportunities; as the market moves from a tightly controlled structure toward a more open and flexible stance, many market participants can more easily take advantage of innovative business models and technological advances such as storage technologies.

This does not mean that energy storage technologies are a panacea or a solution for every market problem; existing power market assets remain the backbone of providing service to customers. However, the normal operation of conventional technologies propagates the inefficiencies inherent in the industry, which often produce the very situations where the flexible capabilities of a storage asset are most needed. In fact, these instances of system instability frequently produce the greatest cost to the system—cost that can be reduced through greater adoption of energy storage technologies. In the previous incarnation of the market, standard regulatory-led practices lent themselves to capital-intensive solutions in meeting these challenges as the redundant equipment could simply be rolled into the rate base. Although functional, this was an inelegant solution to a chronic problem; in an increasingly competitive market, this inefficient solution is no longer acceptable.

For these and other reasons, energy storage technologies provide an option (often the least costly one) that the industry can call on to solve some of its most difficult challenges. Energy storage technologies break the linkage between electricity production and demand, allowing the storage of power for later use. By maintaining a ready-reserve cache of energy, storage technologies can help industry participants overcome such challenges as:

- Improving low utilization of power facilities

- Relieving transmission congestion

- Improving the market potential of renewable energy generation

- Preventing losses from unreliable power quality for end-use consumers

By absorbing or providing power—even a small amount—at precisely the right time and place, energy storage technologies can relieve significant system stress and costs, effectively acting as a *shock absorber* for the industry.

Misperceptions and Realities

Storing energy is a seemingly simple and familiar concept, yet even electric power industry insiders often misunderstand its real potential. The technology is straightforward—storage technologies convert electrical power into chemical, mechanical, or electrical potential energy and retain the ability to reinject it into the power grid when called on. Unfortunately, the stored electrical power is then viewed simply as a static repository of energy, with any value ascribed strictly to its commodity value. This is a simplistic, one-dimensional view of storage technologies—the deliverability of that energy is another key aspect. The ability to decouple the linkage between power production and power demand allows the industry to use storage technologies as both a sink and a source of energy—and both are valuable resources in a dynamic environment such as the electric power market. As a sink, storage facilities can not only absorb energy slowly from generators to optimize their operations, but also absorb rapid surges, preventing imbalances that can affect the power grid's stability. As sources of energy, storage technologies not only can be used to arbitrage energy between off- and on-peak usage, but they can also be used to adjust the rate of energy production and purchases. The ability to affect the supply/demand balance of energy delivered to customers provides the capability to then reduce ramping stress or otherwise optimize the use of generation and transmission assets, or to prevent consumer equipment damage in the retail market from power fluctuations.

Although treated as a new technology, the electric power industry has actually maintained significant storage capability for quite some time now. Therefore, the issue going forward is rather the rate and type of further adoption—not introduction—of these technologies in the market. From the wholesale to the retail market, firms have recognized their flexibility and have invested significant amounts of capital into these technologies for many years. In the wholesale power market, U.S. utilities began in earnest to build a number of pumped-hydro facilities in the 1970s and 1980s, resulting in 20 GW of capacity by the early 1990s (nearly 3% of all U.S. summer capability at the time). Besides their use for arbitraging off-peak power to peak demand, they have proved very useful for providing much-needed stabilization for the transmission system. In the retail market, commercial and industrial firms drive the $7 billion spent globally on uninterruptible power supply (UPS) equipment to provide enhanced power quality and reliability as the demand for better power quality grows.[1] Even the U.S. power industry itself keeps banks of batteries on hand at each of the estimated 100,000 (or more) substations and every major power facility to run the control equipment in the event of a power outage. Existing applications will continue to grow with the market, and new applications will be enabled through expanding capabilities of existing technologies and the continued commercialization of newer ones. As one sign of a maturing storage technology market, some storage technologies now target existing markets already supported by existing battery technologies.

Therefore, energy storage technologies are not simply a solution in search of a problem—they have real uses both now and in the future. In some instances, they enable applications that generate real revenue, whereas in other applications, they either reduce capital expenditures or prevent losses from damage-process interruptions. As these are all valued roles now, no *paradigm shift* in thinking, market structure, or technological basis is needed for these technologies to expand their presence in the market. As the market continues to change and new demands are put on the industry, these existing and new technologies will enable a more flexible way of operating and provide the real optionality managers need. For this reason, the continued introduction of these technologies will not simply be vendor-driven, but actually demanded by the market. In fact, three broad themes at the core of most corporate and public policy goals will promote the continued deployment and use of energy storage technologies:

1. **Efficiency.** Storage technologies can be used to improve the operations and use of generation and transmission facilities, enabling a more flexible and effective market. Even if their capability is short-lived, so too are the times when they are needed most, such as when the wholesale price of power spikes during times of temporary transmission capacity shortage.

2. **Reliability.** Storage technologies can enhance the ability of utilities to provide electricity service to customers without any of the normal or abnormal fluctuations in the quality of the power impacting their customers by improving the power system's stability.

3. **Security.** Storage technologies can provide a means to blunt short-term disruptions on the power grid before they become large-scale problems by giving system operators the fast-response capability provided by a ready-reserve resource— or they can harness additional resources to support restoration of service once a blackout does occur.

Wholesale Power

The wholesale power market continues to prove far more challenging than envisioned by the architects of its transformation, as competitive power prices have brought both new opportunities and new setbacks. Although recent entrants garner much of the market's attention, owners of the existing $500 billion in generating assets struggle to maintain their profitability as the rules in the market change—changes that reveal that previous operating strategies incurred far higher costs than first imagined.[2] These market changes also affect developers of renewable energy projects who find that while many support the idea of developing these domestic resources, integrating them into the market can be far more difficult than first envisioned.

Facility utilization

The move to a competitive power market has stressed the ongoing operations of many existing coal facilities; improving the utilization of these units is one direct way to markedly improve their profitability. Because coal-fired units provide half of all U.S. power generation, improving their value ranks as a public policy concern. As it stands, ramping down during off-peak hours (nights, weekends) to match the current load leaves these units with appreciable unused capacity. Although generally termed baseload, coal-fired units do not enjoy the must-run status of nuclear units, leaving the fleet average of all coal units around 70%, whereas nuclear units have been able to increase their levels well above 90% (fig. 1–1). Increasing the utilization of these facilities would lower average operating costs—not only increasing overall revenue, but also the profitability of these facilities. The off-peak production cost of power could also benefit (from higher plant utilization), as some market studies point to increases in the heat rate of individual units of anywhere from 5% to 20% when running at low power.[3] Anecdotal evidence even points to some older units whose total operating costs reportedly increase upwards of 40% when operating at half power. Beyond lower operating costs, per-kWh-hour emission levels would also decline, as the power facility is able to operate more closely to full power. Although this overall improvement may not be dramatic, any improvement on this issue in an era of ever-tightening environmental restrictions could prove beneficial.

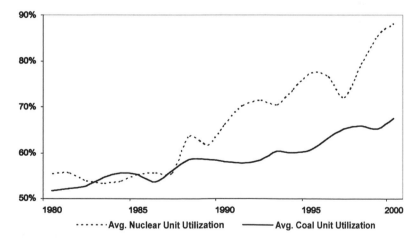

Fig. 1–1. Coal units lag nuclear in utilization improvements (Data: U.S. Department of Energy).

Large-scale energy storage facilities can help increase the utilization of coal facilities—or other units with excess capacity—by acting as an energy sink during energy-usage off-peak periods. Coal facilities with the most to gain would be those with significant excess production capacity as they should have the lowest marginal cost for additional power production. Existing large-scale energy storage facilities (pumped-hydro, compressed air energy storage [CAES]) already operate in this arbitrage role today, and even offset the need for some additional peaking capacity during peak demand. Producing additional power at night can even help daytime air quality, especially during hot summer days prone to ozone warnings, because the emission levels of a CAES facility (and especially those of a pumped-hydro unit) are far lower than those of a peaking combustion turbine unit.

Cycling damage

Matching the changing daily load can also have negative impacts on mid-merit power facilities, and to a far greater degree than previously thought. Recent market strategies have evolved so that older fossil-steam (and even some combined-cycle) units in the mid-merit role are forced to cycle with the load, resulting in many facilities cycling heavily every day—with some even cycling on and off daily . What is becoming better understood is that the cost previously ascribed (and used currently in dispatching decisions) to these activities is significantly below their real value; the true cost of this cycling is much higher—sometimes more than 10 times the previously used value—making the current dispatch decisions far from optimal. One of the leaders in the field of power plant failure/performance analysis is APTECH Engineering Services, which has spent many years evaluating these problems. Based on analysis of real operating data from more than 200 power generating facilities, Steven Lefton, vice president of the firm, notes that if the true cost of cycling was used, an average utility power system could reduce its total production costs by 5% (and increase profitability significantly). This heavy cycling and rapid load following is taking its toll by damaging plant equipment and requiring far larger maintenance costs. Although the older fossil units have proved far more rugged than first thought, the resulting number of warm or even cold starts is far more than first envisioned for these units, especially as some are now older than many of their operators.

Even combined-cycle units—having improved significantly from their first entrance in the market—still find the cycling troublesome. The resulting effect on these units is that the wear and tear is accelerating, resulting in higher operation and maintenance (O&M) costs, longer and more frequent forced outages, higher heat rates, and shorter life expectancies for critical plant components (fig. 1–2).

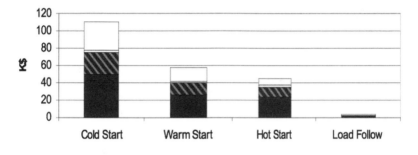

■ Maintenance, Operation & Capital Cost   ◪ Forced Outage Replacement Power Cost
□ Low & Variable Load Extra Fuel Cost   ▥ Startup Fuel, Aux Power, etc. Cost

Fig. 1–2. Power generation cycling cost breakdown (Courtesy of Aptech Engineering Services, Inc.)

Large-scale storage facilities can help these mid-merit power facilities in two ways. First, they can provide a sink for energy at night sufficient to keep these plants from shutting down, precluding many of the excessive warm starts that create the costly and cumulative damage to the units. Second, the storage facilities can provide faster-responding load following services so that the demand on power generation facilities is not as dramatic during times of rapidly changing demand, allowing these older units (and combined-cycle facilities) to ramp along their optimal design path rather than what the market demands. Both of these roles would reduce the cyclical wear on the power facility equipment significantly, greatly extending its life and lowering the maintenance replacement costs. Extending their analysis to include this application of energy storage,

APTECH Engineering Services estimates that including storage into the mix can improve the performance of a large utility system by anywhere from 10% to 28% from reduced cycling damage and dynamic heat rate effects.[4]

Renewable energy

Increasing the use of renewable resources such as wind is one of the prime goals of U.S. energy policy makers. Not only does wind provide a domestic source of power, it is an environmentally clean source of energy. Wind power in particular has seen dramatic growth, with a total of more than 6,000 MW of wind turbines installed by the end of 2003. Far more is possible, with the American Wind Energy Association (AWEA) estimating that the United States alone has three times the wind power potential compared to its current electricity usage today. However, these new resources face market penetration challenges as they become integrated into the power market. Wind energy production potential is generally noncoincident with peak demand periods; most (roughly two-thirds) of the energy that can be produced from wind turbines is outside of the weekly peak demand (and pricing) periods. In addition, wind resources are variable—making the stable delivery of power from the wind turbines hard to achieve. Most wind resources developed to date have been in remote locations where the existing utility transmission capabilities are generally limited.

Advancements in wind turbine technology have created many new opportunities for even more development, but challenges remain. Although modern wind turbines can now produce power competitively with natural gas combined cycle units, there is a growing realization that the wind energy potential in many areas will remain stymied because of the above-mentioned challenges. Noncoincident production means that much of the power is only sold at off-peak prices. Unpredictable power means that not only is the capacity payment heavily discounted, additional ancillary services are needed to integrate the wind power into the power grid—again lowering the value of the wind energy to the system operators. Much effort (especially in Europe) has gone into extending weather forecasts into *wind-power* forecasts, but weather forecasts have their limits, especially in the time frame required by the power industry. Distant resources mean additional costs for transmission or even the potential loss of sale because the wind energy could be constrained off the power grid during peak demand. Therefore, without addressing these attributes of the wind resource, modern wind turbines may very well be able to turn vast quantities of the nation's wind resources into electrical power, but that wind energy will continue to be penalized or even unusable in the market.

Storage can improve the value of wind energy (and reduce the project risk) by both increasing the value of the wind energy and reducing the current discounting of its output value. At its heart, storage offers a means to decouple the production of wind energy from demand, and to provide that power in a dispatchable and stable manner to the market. These capabilities can directly mitigate the negative aspects of wind resources to improve the value of the wind energy and promote a greater market penetration of wind power. For instance, on small or isolated power grids, wind resources are plentiful, yet the variability of that potential resource could mean additional generating units must be added to stabilize the power grid, actually increasing the overall cost of power. A growing number of wind and storage projects with short paybacks are proving the success of this strategy today. With storage, these small power grids can rely on wind power to a far greater degree, reducing or even eliminating the need for the diesel generator and its expensive fuel needs. In the larger wholesale power market, a number of wind and storage projects are currently being evaluated. Although some will provide a *capacity-firming* capability to support the wind turbine output, others plan to optimize the design and use of supporting components such as the transmission facilities. Storage can even help ensure that distant wind resources are able to get their output to market if the transmission capability is severely limited, threatening to strand the renewable energy far from where it is wanted.

Transmission and Distribution

The transmission and distribution sector is arguably the area most in need of attention in the power industry. Developed with an emphasis on reliability rather than efficiency, the system was built out to ensure sufficient capacity during peak demand periods—leaving much of the system largely unused during off-peak times and producing an average system utilization that rarely surpasses 60%. The $100 billion and $250 billion worth of assets in the transmission and distribution markets, respectively, make it apparent why it is necessary to improve their capability and cost effectiveness.[5] Although the industry is attempting to transform the transmission system into a real national grid with self-correcting market controls, the distribution networks struggle to satisfy consumer demand for ever-improving levels of power quality and reliability. Confounding the plans for these changes are falling infrastructure investments, helping to cause growing congestion problems during peak periods as strategies to improve and standardize the rules for operating the system remain slow in coming.

Infrastructure underinvestment

Underinvestment in the underlying infrastructure of the nation's power grid is a well-known and growing problem. Reasons for this lack of sufficient investment are a bit circumspect as they include historical construction patterns, regulatory risk, and poor financial performance.

It should be noted first that the system was built around a central tenet that electricity can only be produced when it is needed, and it must be used once it is produced. Operating the power grid as simply one big just-in-time delivery system in this way, however, is extremely wasteful. It requires significant capital to be tied up in a production and delivery system that must always be capable of meeting the highest expected demand, but the highest expected demand is only required infrequently—for much of the year, the system remains significantly underused. In addition, economic necessity and technical standards have curbed development and installation of large upgrades or additions to the system, producing the familiar stair-step expansion pattern for transmission capacity. For example, as demand grew rapidly throughout the 1950s and 1970s, significant transmission investment took place to provide the wide safety margins common in those days. After the energy shocks and recession of the early 1980s, however, lower demand growth permitted existing assets to meet demand longer than previously expected, allowing transmission expansion budgets to be pared back. As increasing regulatory changes fueled the uncertainty as to what would be the final structure of the industry, investment continued to taper off as utilities looked for markets with higher returns and that were perceived as less risky. Unfortunately for the transmission market, these regulatory changes favored the generation sector throughout the 1990s, making that market a far more attractive investment opportunity and thus reinforcing the decline in transmission investment, which continues today.

As investments in the infrastructure continue to fall behind the need for them, raising the efficiency of the existing assets becomes even more important. Although the U.S. Department of Energy (DOE) estimates that the high-voltage transmission network is comprised of assets worth $100 billion, the Edison Electric Institute (EEI) estimates that the industry spends only $3 billion annually on maintenance, and $2 to

$2.5 billion for capacity expansion (230 kV and above). In fact, over the past 25 years, the group has noted that investment in the transmission system has declined by $115 million per year, leading to the present state, where annual investment (in nominal dollars) has reached half of what it was in 1975, whereas actual kWh sales have doubled. Without additional investment, badly needed upgrades will continue to go undone, further stressing existing equipment already operating past its expected life span and making it much more susceptible to a breakdown. Although the EEI estimates that an additional $56 billion will need to be spent to catch up with the legacy of underinvestment, the problem continues to grow as the DOE expects demand to grow 22% from 2000 through 2010, whereas the North American Electric Reliability Council (NERC) is only estimating the expansion of the transmission power grid by 5%.[6]

Although many hold out for a regulatory-led movement to increase the level of investment in the industry's transmission infrastructure, storage can help increase the usefulness of existing assets—a far more effective means of investment for utilities in today's environment. By making existing equipment capable of meeting future demand and postponing otherwise needed upgrades (especially in lower-voltage systems where the investment problem is even more acute), the increased use of these assets will help improve the financial performance for their owners. In theory, the case for a storage component in the power industry would therefore seem obvious. With a storage component, the owners and market coordinators must build out only what is necessary to carry a heavy, but more typical load—resulting in a much higher utilization of their existing equipment, and the system overall. Even a few installations of storage technologies could provide badly needed supplemental capacity (bulk energy and system stability) during peak demand, lowering capital costs and mitigating supply disruptions. At the transmission level, storage technologies are already a cheaper alternative for Wisconsin Public Utilities than existing voltage stability equipment or capacity expansion alternatives. At the distribution level, other utilities such as PacifiCorp are beginning to use other storage technologies to postpone the need for upgrading remote power lines, which is a constant challenge that is both costly and lengthy.

Transmission congestion

One result of the lack of investment has been the increasing level of congestion because of a deficiency of excess transmission capacity during peak demand periods (fig. 1–3). Congestion is a serious problem for the transmission market, affecting large areas of the power grid when vital links such as the infamous Path-15 in California become oversubscribed, contributing to system instability and more volatile power prices. As this condition worsened through the later the 1990s, independent system operator (ISO)–levied congestion charges amounting to hundreds of millions of dollars per year were passed on to customers. The total bill for congestion charges is now measured in the billions, with more charges expected in the foreseeable future. Besides direct monetary costs, congestion also affects planning for power project development, such as where to locate; while some target load pockets where prices may be high, others such as wind developers are penalized because their location is often predetermined, and the question becomes whether they can get the power to market. In fact, some wind farm developers do not build out their site to its fullest extent because of the lack of transmission capacity during peak times. Building out vast excesses of transmission capacity is difficult (beyond cost, simply permitting a new high-voltage line today requires Zen-like patience), so other, more realistic solutions must be found to raise the carrying capacity of the existing network. Ongoing efforts to raise the carrying capacity of the transmission network through more efficient operation, expanding existing right-of-ways, and using flexible AC transmission systems

(FACTS) devices are being promoted, particularly in susceptible areas of the power grid. By improving the carrying capacity at these bottlenecks, the entire carrying capacity of the power grid can be improved.

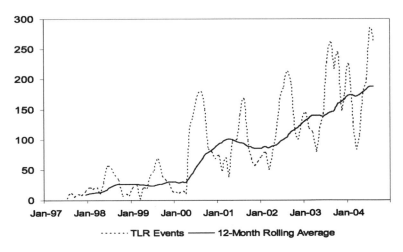

Fig. 1–3. Transmission load relief incidents on the rise (Data: NERC).

Congestion issues also extend into the distribution market. The previously mentioned underinvestment grows as one moves down in transmission size, leaving the system stressed during peak demand. Because of its geometry, congestion at this level can produce serious power quality and reliability issues for customers in highly localized areas by preventing power from reaching end users; long-distance, low-voltage distribution lines are especially prone to voltage problems for this reason. Demand on these lines is extremely variable. Therefore, even if the average utilization of the line is low, there can be times when demand exceeds the transmission capacity of the power line. In fact, only a few instances of maximum usage (or near) per month are enough to warrant an upgrade in transmission capacity to alleviate the problem. When this occurs, it is common for the line to be upgraded significantly to postpone the next required upgrade, ensuring that the average utilization of the power line will remain low.

Storage technologies can help alleviate both the lack of transmission capacity and the stability problems that ensue during periods of high demand. Sited past a bottleneck, they can provide a prepositioned source of energy, which will allow the system to *ride through* a few short-term peak demands during the month—postponing the need for an expensive upgrade on the line until a more sustained level of demand warrants it. With accompanying power electronics, they can also provide a means to maintain voltage stability at these vulnerable locations—although the act of providing real power in these stressed areas can also improve the system's voltage stability. Although there are alternative means to supply short-term power (small peaking facility) or voltage stability (capacitor banks or FACTS equipment), neither of these choices is as flexible as an energy storage facility. Many storage technologies developed for this role are also mobile, so they can be moved as the conditions and needs of the power grid change.

Ancillary services

Ancillary services are essential to maintain the stability and reliability of the power grid for the transmission of electric power; unfortunately, they remain mostly misunderstood—even by many within the industry. Frequency regulation, voltage control, and contingency reserves are just a few of the components that must be monitored and constantly regulated because changing load conditions will interact with the geometry of the system to affect the reliability and deliverability of power to customers. As power generators provide many of these services, the move toward divestitures (and a focus on peak power sales) has left providers with less incentive to provide these vital support functions just as the need is growing. Once simply a concern of utility system operators, the Federal Energy Regulatory Commission's (FERC) desire to create a more self-correcting electric power market led it to create a competitive market for some of these services coordinated at the ISO level. This unbundling of services is modeled after its previous successful actions in the natural gas market, where substantial growth in flexible contract sales occurred once the transmission charges were unbundled. Although more work is needed in the power industry, this unbundling has finally provided the beginnings of much-needed price visibility of these ancillary services. Because the provision of these ancillary services is such a central component to the working of the wholesale market, it would be impossible to think that the wholesale market could operate competitively and successfully without real reform of these functions.

Most well-regarded analyses of the U.S. ancillary services markets originate from work done by Eric Hirst and Brendan Kirby. In a 1998 study, they estimated that the United States spent $12 billion on ancillary services in 1996—fully 10% of direct energy costs. Other researchers have found similar (or higher) costs. Most assumptions point to a market that has grown significantly since then, as many of these values represented embedded costs from an era prior to deregulation and not opportunity costs in a competitive market. Specifically, many of these ancillary services are linked to the spot price of power, or simply the scale of the overall power market—both of which have grown significantly since the mid-1990s. However, simply using Hirst and Kirby's original $4.15 per MWh metric for all ancillary services, today's total ancillary service costs amount to more than $15 billion annually; with market-derived prices for these services, the total market value could easily be far larger.[7]

A handful of storage technologies is already providing ancillary services to the power market, with others set to begin competing in the near future. One core strength of these storage facilities is their responsiveness. Because the role of these ancillary services is to counteract rapid variations in power flows, responsiveness has real value for these applications. For example, CAES facilities have the ability to come to full power within 10 minutes (and pumped-hydro facilities can respond within even less time), whereas natural gas turbines can require 25 minutes, and combined cycles 45 minutes in some instances. Coupled with a large storage reservoir, energy storage facilities can provide load-following or other longer-term energy supply ancillary services. Smaller storage technologies can also support the stability needs of the power grid. Placed at strategic locations, these units can directly reduce the stress at weak points on the power grid by providing both rapid energy balancing and voltage control. Some of these units can be moved as the stability conditions of the power grid change. In the near future, more storage technologies are targeting a role in this market as well. For instance, a MW-scale, high-cycle capability storage system could provide frequency regulation services by absorbing and discharging essentially the same energy far more quickly than a power facility could respond. Although not to the same scale as a power facility, the much-smaller storage facility could, nonetheless, offer real value by providing support to troubled areas of the power grid.

Retail

Energy use is of growing importance to many commercial and industrial customers—but not by choice. According to the DOE, commercial and industrial firms spent $133 billion for electricity service in 2000; as the U.S. economy continues to grow and to increase the penetration of electrical equipment, their electrical demand is expected to be 20% higher by 2010. Although most firms will not be fully exposed to volatile time-of-use rates over the next few years, some are already having to create strategies not only to save money on their electric bills, but, more importantly, to prevent problems with electric power quality negatively impacting ongoing operations. For the future, most customers understand that the end result of restructuring will be that they will be far more responsible for looking after their own best interests than in the past.

Power quality

Poor power quality is a growing concern to commercial and industrial firms. This is because poor power quality is quickly becoming a hidden drag that threatens to erode much of these firms' hard-won operational improvements. It is not a decline in the quality of delivered power, however, that is the culprit here; it is, rather, the increasing expectations customers have for their electricity service. Whereas industrial customers fear even transient power fluctuations that can disrupt their high-speed processes, commercial firms worry about momentary power sags as fault-intolerant information technology spreads throughout the firm. Beyond simply causing a loss of in-process work, larger power-quality events can even damage this increasingly expensive equipment. Surprisingly, most of these power quality disturbances are short-term, according to the Electric Power Research Institute (EPRI), with 98% lasting less than 30 seconds and 90% lasting less than 2 seconds.[8] What these firms want—either from their service provider, or, increasingly, from their own energy management strategy—is a means to essentially provide a capability for *loss prevention* from these transient frequency variations or voltage surges and sags. The sad fact is that the real impact of these disruptions will continue to grow, however, as most firms have "wrung out" any savings in their operations, and any interruptions in the resulting schedule are increasingly expensive.

Estimates vary, but the annual cost to the U.S. economy from poor power quality currently ranges from $119 to $188 billion (EPRI) to $150 billion (DOE) from interruptions in operations, lost work time, and damage to increasingly expensive equipment.[9,10] Because of the magnitude of these losses, which stymie economic growth in hard-hit areas—and their ever-upward direction—it is no wonder that the DOE has raised this issue as a central part of its national energy policy. To put these values in the perspective of an individual firm, one survey from a UPS manufacturer estimates that an average manufacturing facility loses $1 million per hour during a power interruption. For some firms, such as electronics manufacturers, these costs are even higher, with estimates from the DOE putting even a five-minute unexpected shutdown at some chip-fabrication facilities at more than $5 million from product loss and restarting the process.[11] Information technology poses even a larger problem, where literally billions of dollars of activity can be affected within that first hour after a shutdown at a bank or other financial-transaction firm.

Because of this impact on the bottom line, many commercial and industrial firms already use energy storage technologies to protect critical loads from poor power quality. Made up of power electronics and an energy storage component, these UPS systems are designed to protect electrical equipment from momentary but potentially damaging power sags and outages. As the impact of poor power quality grows, the global market for UPS systems was estimated to be $7 billion in 2002, with growth over the foreseeable future to average 7% per year.[12] With sufficient power, some of these UPS systems are also able to act as a ride-through or bridging-power source for essential equipment in the event of a power failure. Generally, 15 to 20 seconds are needed to either bring online a backup power generator or switch to a different local circuit. Maintaining a backup generator on-site remains a popular strategy to ensure power for critical loads, as evidenced by the multibillion dollar market for these systems. However, many firms requiring utility-comparable power quality (or better) from their on-site power units have found their capability wanting—especially if the load remains variable—because the smaller on-site generators will have a far more difficult time in responding to any changing load than the utility's system power would have. Here too, newer storage technologies such as flywheels show promise in providing frequency regulation to enhance the quality of power when the facility is isolated from the power grid.

Cost of energy

All commercial and industrial firms would like to spend less money on their energy costs; in fact, the desire of these firms to lower their energy expenditures has been one of the core driving forces behind retail restructuring. The importance of reducing energy costs has created new strategies—such as the growing use of adjustable speed drives to replace older, more energy-intensive fixed-speed motors. Now, however, these firms' original desire for simply lower utility bills has been replaced by the specter of utility bills rising unpredictably based on activity outside of their control in an increasingly complex retail power market. For many firms, even a small rise in energy costs could wipe out their hard-won cost reductions in operational efficiency, so many of the variable expenses have been put (again) under review to look for savings. Many find that both the commodity charge and the demand charge contribute to the variability in the total amount a firm spends on electricity service. Searching for a better means to manage their energy needs, many are now concentrating on power-intensive activities that occur during peak periods, looking to reduce the demand charge to lower their variable energy costs.

For some firms, the cost of energy is a central component of costs, warranting the development of an involved energy management policy; for others, the effort can quickly cost more than the potential savings. For example, in energy-intensive industries such as the aluminum industry, fully one-third of the cost of production is electricity. However, the cost of electricity generally represents only 2% to 3% of the total cost of most commercial and industrial firms, and only 1% for most office buildings. Surprisingly, even for many of these firms, where the overall cost would not seem to warrant any action, what is commonly found is that certain components of their energy usage may be suitable for incorporating energy storage technology into part of their overall load. For instance, the DOE notes that 25% of commercial usage for electricity is for cooling load, a highly cyclical operation. Although heavy energy use increases the commodity energy charge, highly cyclical loads such as cooling or batch processing can increase the demand charge component of the bill dramatically. Currently, roughly 30% of all firms are subject to demand charges; if much of the reason for this stems from a highly variable load such as cooling, then many times it may make sense to devise a means to eliminate that daily surge in demand. By reducing the volatility in their energy usage, the firm's demand charges will decline, and total costs will be far more predictable—reducing significant amounts of uncertainty in their potential earnings.

Energy storage technologies can be—and already are—used to lower companies' power bills. Obviously, firms with high-priced tariffs stand to benefit from storing low-cost power at night and using it during peak demand the next day. However, many others (even those where energy costs make up a small portion of costs) can also reduce their cost of electricity if one part of their load is highly variable—for instance the previously mentioned cooling load. Here, thermal energy storage is already used by many commercial firms to essentially create large blocks of ice at night, which are then used to assist with the daily air-cooling load for air conditioning. According to some vendors, these units can many times reduce cooling peak power demand by upwards of 50%, producing a 30% overall reduction in the cost for cooling. If integrated into the design of a new building, these systems can even result in a reduction of the cooling infrastructure equipment size requirement, sometimes by 40% to 60%—providing additional benefits for the firm. Because of these results, typical payback periods for these installations (new and retrofit) can be realized in one to three years.[13] Newer storage technologies are enabling a wide variety of other useful applications as well. For instance, flywheels are capable of capturing wasted energy in repetitive motion situations. This role of acting as a dynamic sink and source for power fits well with transportation sector applications, where repetitive starts and stops produce very inefficient use of energy. Although container port lifting cranes and light-rail/subway systems all exhibit these usage patterns, they are currently under served because previous battery technologies could not support these applications, leaving these firms with no solution to their problems. If an energy storage solution is used to lower the operator's costs, even the local utility stands to save because the load swings on the power lines feeding these installations are reduced.

References

1. *World UPS markets: Alternative energy storage solutions.* 2003. San Jose, CA: Frost & Sullivan.

2. U.S. Department of Energy. 2003. **Grid 2030**—*A national vision for electricity's second 100 years*, 3. Washington, DC: U.S. Department of Energy.

3. Grimsrud, P., S. Lefton, and P. Besuner. 2004. *Energy storage systems (ESS) provide significant added value by reducing the cycling costs of conventional generation.* Sunnyvale, CA: APTECH Engineering.

4. Ibid.

5. U.S. Department of Energy. 2003. **Grid 2030**—*A national vision for electricity's second 100 years*, 3. Washington, DC: U.S. Department of Energy.

6. Edison Electric Institute. 2002. *Energy infrastructure: Electricity transmission lines.* Washington, DC: Edison Electric Institute.

7. Hirst, E. and B. Kirby. 1998. *Unbundling generation and transmission services for competitive electricity markets* (ORNL/CON-454). Oak Ridge, TN: Oak Ridge National Laboratory, U.S. Department of Energy.

8. *EPRI distribution system power quality monitoring project (DPQ study).* 2003. Palo Alto, CA: Electric Power Research Institute; Knoxville, TN: Electrotek Concepts.

9. Electric Power Research Institute. 2003. *Electricity technology road map: 2003 summary and synthesis*, 14. Palo Alto, CA: Electric Power Research Institute.

10. Swaminathan, S., and R. Sen. 1998. *Review of power quality applications of energy storage systems* (SAND98–1513). Albuquerque, NM: Sandia National Laboratories.

11. Gyuk, I. *Electrical energy storage* (presentation). Electricity Storage Association, 2000 Summer Meeting. Seattle, WA.

12. *World UPS markets: Alternative energy storage solutions.* 2003. San Jose, CA: Frost & Sullivan.

13. MacCracken, M. 1993. Thermal energy storage myths. *ASHRAE Journal.* 45 (9): 36–42.

2 STORAGE IN OTHER ENERGY MARKETS

Storage is a fundamental component of the petroleum, coal, and natural gas markets. As these industries transitioned from highly regulated, commodity delivery networks (creating logistical issues) to highly competitive, deregulated markets (fostering hedging/speculative strategies), storage has been there—first enabling the transition, and then becoming a key provider of services. Those involved with public policy development in the electric power industry should review these other markets' experiences to learn how storage can become another tool in promoting a better working power market. In these other energy markets, the introduction and use of storage has undergone a surprisingly similar learning curve. First recognized for leveling variations in physical supply, a more integrated use of storage allowed lower-cost transmission upgrades and system expansion costs. Deregulation in these markets—with particular respect to natural gas—leveraged the ability of storage facilities to provide ready-reserves of the commodity to create greater customer choice and, surprisingly, a greater demand for the use of storage facilities. As wholesale physical trading begat even greater financial trading, storage assets served not only as the basis for many of these contracts, but also as a tacit physical backstop for the trading activity. In fact, it was the availability of storage facilities in the natural gas industry that allowed the overall market to weather the restructuring of that market from deregulation.

Because they represent a physical hedge, storage facilities are most useful in wholesale commodity markets. By improving the service past a bottleneck in a network through the use of a prepositioned cache, or arbitraging between markets with uneven service (improving deliverability of the resource to take advantage of the uneven local prices), physical and price volatility can be more easily mitigated. Storage facilities also have a role as the market matures. As a market clearing price and forward price curve develop through greater trade, storage facilities act as the basis or physical backstop for trading—first physical and then purely financial trading. An interesting point is that although storage facilities need wholesale markets to reach their full values, wholesale markets in turn need storage facilities to operate properly and facilitate lower end-use costs and additional customer choices.

As the electric power industry finds itself in the midst of its own fundamental change, significant lessons can be learned from these other energy markets in how storage can be usefully applied. However, because electricity is not simply a commodity but a service where moment-to-moment quality distinctions are important, there are real applications for storage in all three power market segments: wholesale, transmission and distribution, and retail. Lessons learned from the petroleum, coal, and natural gas markets can provide significant guidance in how storage can act as a substitute for transmission upgrades or to improve customer choice. The electric power industry already uses storage to improve customer service throughout the value/supply chain, although this practice receives little recognition. Unlike natural gas storage, however, energy storage in the power industry has not seen a similar level of public policy support because of different market structures and competing technologies. However, maturing storage technologies and changing market conditions in the power industry are now providing an opportunity for expansion of energy storage for use in a variety of applications throughout the industry. Even if the raw quantity of commodity storage in the power industry may never equal that in other industries, the speed of the power market provides for the opportunity to have as great or even a greater overall impact in service quality. For that reason, the question may not be why there should be a storage component to the electric power industry, but, rather, why not?

Petroleum Market

The petroleum industry is one of the largest industries on the planet. Not only does it literally span the globe, it is also one of the most mature and complex delivery markets of any kind. As the industry has grown (from the U.S. perspective) from a national one to one where more than half of all petroleum comes from overseas, the use of storage has increased from improving the flow of petroleum in interstate pipelines to becoming the basis for more than $600 billion in trading that is now only loosely based on physical delivery.

The petroleum industry and the need for storage

Wholesale petroleum market. Americans in the United States spent more than $87 billion on petroleum products in 2002. To deliver the 15 million plus barrels per day (bpd) that were consumed, the 6 million bpd of domestic production were supported by 9 million bpd of imported oil.[1] This growing reliance on imported oil has been an area of concern for many who fear a supply disruption. Although efforts to find additional domestic resources continue, significant effort is underway to streamline the delivery networks to speed the transit time of product, while providing redundancy to the system so localized disruptions do not affect larger areas.

The need for storage. Petroleum stocks are used primarily to keep crude oil and refined products moving efficiently from wellhead to end users. As an immediate or ready-reserve source of supply, petroleum stocks provide a physical backstop against both expected and unexpected market disruptions. For instance, in production regions stocks allow for production to be disconnected from pipeline supply; otherwise, well output must be controlled to match demand, significantly affecting the life (and total output) of the well. Along the pipeline, storage provides an essential balancing role in case of a shortfall or in case regional demands change. However, these disruptions while the petroleum is moving are essentially logistical issues because of the speed of the crude oil. On average, crude oil and products move through a pipeline at 3 to 8 miles per hour, and crude oil tankers travel at 16 knots, requiring 12 to 18 days for product to travel from Texas to New York, or 30 days for crude to reach the United States from the Middle East.

Historically, the first storage of oil (stocks) simply consisted of large wooden boxes or even simply large pits in the production regions. However, because crude oil and petroleum products are volatile and evaporate easily, storage techniques to contain these vapors began to improve after the U.S. Civil War (1865), when riveted wrought iron tanks began replacing the wooden variety. This style of storage facility remained the primary way to maintain oil stocks for the next 50 years. Advancements in welding technology in the 1920s led to another round of improvements as steel storage tanks were introduced as more durable storage facilities. Throughout all of this period, the sheer volume of oil stocks continued to increase in step with production, with relatively high levels required because of the poor transmission coordination capability of these far-flung enterprises. As oil demand grew quickly following World War II, stocks of petroleum products increased to maintain deliverability in the ever-growing network of pipelines—and increasingly import terminals. With the greater reliance on imported oil came more supply instability stemming from political instability in foreign-production regions, increasing the need for additional supplies to mitigate temporary supply disruptions. After the early 1980s, however, stocks of petroleum products began to decline as efficiency improvements reduced the need for excessive amounts, and increasingly volatile prices increased the carrying costs for commodities.

Regulatory impact on storage. Besides health and safety regulations, the only regulatory development concerning storage has been the establishment of agreements by International Energy Agency (IEA) member countries to hold stocks of crude oil to mitigate supply disruptions, which are often politically motivated. With the oil-sharing rules of IEA, member countries share the burden of an oil supply shortage, creating some resilience to the demand component of the global supply network. These agreements require that each participating nation hold stocks equal to 90 days of imports. Most member countries meet the requirement with industry-owned stocks that can be commandeered in an emergency; however, the United States, Japan, and a few other nations also hold government-owned stockpiles. These 90 days of separately held stocks were estimated to be worth around $100 billion in 2004. Crude oil stocks owned by the U.S. government are held in the Strategic Petroleum Reserve (SPR), originated in 1975 as a temporary replacement for imports during significant supply disruptions.

Facility types and operations

Storage tanks fall within three levels that correspond to their position or level within the supply chain:

1. **Primary.** Facilities are used for crude oil storage, refinery operations, and petroleum products storage at wholesale marketing facilities with more than 50,000 barrels of capacity. Above-ground tanks are the most common type of primary storage facility in the United States, with the average facility holding in excess of 500,000 barrels.

2. **Secondary.** Facilities include wholesale marketing installations with a capacity of less than 50,000 barrels and retail establishments. These facilities balance the distribution demands of tanker trucks delivering products to retail markets as the demand fluctuates because of seasonal and economic factors.

3. **Tertiary.** Storage comprises end-use consumers' home-heating fuel tanks and vehicle fuel tanks.

Market uses and economics

An important role of oil stocks is as an indicator of spot price movement. Normal fluctuations in crude oil stocks occur as the market participants attempt to balance seasonal changes in the demand for refined products and the projected supply of crude oil. These discretionary stocks also effectively serve as a physical hedge against price disruptions in the financial trading market. With 10% to 15% of the world's primary petroleum stocks discretionary, movements in this small subset of oil stocks are followed closely.

In the short term, stocks act as a real-time indicator of the relative availability of deliverability. An inventory increase or decrease that appears small may actually result in a large price change because this movement can actually represent an important shift in the usable volume of inventories. Because of the depth and competitive nature of the petroleum market, the forward price curve causes a feedback impact on stock levels, and together they act as a self-correcting agent for the market in general. For instance, when stocks are low, spot market purchases increase, driving up prices, which begets more production and finally an increase in stock levels. When stock levels reach points higher than historical norms, spot purchases decline, begetting lower production and encouraging limited additional stock fills. Besides the spot trading market, stocks of petroleum products influence refiners' day-to-day operational decisions in the short term. By balancing the cost of storage with other issues, such as refinery margins and expected prices for certain products, refiners attempt to maintain sufficient stocks for operational needs while boosting profitability. Outside of the refineries, storage of particular products is also used to bolster the supply of a product during its demand season to assist with optimizing refinery usage.

In the long term, stock levels are governed primarily by the carrying cost of storage, which is made up of two parts: the actual cost to store the petroleum and the opportunity cost of ownership. The physical storage cost can vary significantly over time, depending on the type of oil (crude or product) being stored, the current availability of storage, if the storage capacity is owned or has to be rented, the price of the oil, and so forth. The opportunity cost is then the implicit cost of working capital tied up in owning the oil during storage. For instance, according to estimates from the Department of Energy (DOE) based on average prices in the first half of the 1990s, holding crude oil for a year would cost a company approximately $1.50 per barrel if it had its own storage and $4 per barrel if it had to rent storage tank space.[2] In this way, the cost of storage can be important to the overall strategy of the firm, even if it is not widely evident.

Coal Market

Power generation derived from coal provides more than 50% of all electric power in the United States, and coal is expected to remain the single most important fuel to the electric power industry for the foreseeable future. Moving this fuel from mines to major power facilities, both of which number in the hundreds, across sometimes thousands of miles, drove the need for a flexible storage solution capable of maintaining continued operation through sometimes-severe supply disruptions.

The coal industry and the need for storage

Wholesale coal market. Slightly more than 1 billion tons of coal are mined each year in the United States, with around 90% of that destined for the power sector (the remainder is for steel making). Coal comes in a wide variety of heat content types and contains varying amounts of impurities, depending on the seam of origin. These impurities—sulfur, moisture, ash, and others—greatly affect where the coal can be used, requiring many power facilities throughout the United States to source coal from thousands of miles away to obtain the coal appropriate for the power plant.

During the 1970s and 1980s, many utilities—responding to expected shortfalls of natural gas supplies, problems with nuclear power facility building programs, and high inflation—entered into long-term coal supply contracts of sometimes 20 years, or more. The fears of insufficient natural gas supplies for the power industry, in addition to industrial and consumer use, restricted the construction of power plants that use petroleum or natural gas as their primary fuels, and encouraged the building of many new coal-fired facilities. The move to longer-term contracts was considered a wise one at the time because there were no significant new coal resources—or improved mining techniques—expected to increase the supply of cheap coal. After most of these contracts were signed, however, new underground mining techniques were developed and enhanced, along with vast surface mining operations in the western United States (especially in Wyoming's Powder River Basin), which drove down the cost of coal production dramatically. As the long-term contracts began to expire in the 1990s, many power generators were able to move to shorter and far more flexible contract terms, and a greater reliance on spot purchases. Many coal customers have, therefore, been free to experiment with using a wider variety of coals, such as the low-sulfur coal from Wyoming's Powder River basin, which now produces 33% of all U.S. coal, up by more than one-half of its output from only 10 years ago. With a significantly more flexible and shorter-term delivery focus, many power facilities now require a far more detailed logistical strategy for handling and storing coal.

The need for storage. In the 1950s and 1960s, many of the new power facilities were powered by petroleum and natural gas because the high price of delivered coal—governed by high mining and transportation costs—drove away new plant orders. To provide more competitive prices, the coal-mining industry worked with the railroads to create longer *unit-trains* of 100 cars or more that could provide lower transportation rates—resulting in lower delivered costs. Each of the 100 cars carried 100 tons of coal (125 tons in newer cars), so a unit train could move 10,000 tons of coal or more at once. However, to take advantage of these special rates, the train needed to be loaded and unloaded in four hours or less (far more quickly than the sometimes day-long events common before) to meet more demanding scheduling requirements. These larger deliveries, therefore, resulted in the need for larger, more flexible storage facilities at the mine mouth and at the power stations. In tandem with this transportation evolution, new coal-fired power plants continued to increase in size as developers attempted to reduce the cost of producing power—also requiring significantly larger storage facilities to hold and handle the fuel for these behemoths. Some power facilities contained individual boilers rated upwards of 600 MW each—up by nearly an order of magnitude from the previous generation of power facilities.

Regulatory impact on storage. Whereas the coal mining industry operates under an extensive regulatory regime, no regulatory changes have targeted—or are expected to target in the near future—coal storage activities directly.

Facility types and operations

Coal storage is primarily used at both the mine mouth and at the power facility. These storage facilities come in a variety of shapes and sizes, ranging from simply open storage piles to enclosed bins and silos, which is determined by the need to balance maintaining the maximum amount of coal in one place with the inherent problems of storing that much coal in one place—namely dust control, oxidation, or added moisture from the environment that can cause spontaneous combustion or degradation of the quality of the coal. These piles can become quite large; for example, a pile 70 feet high with a 200-foot base can hold upwards of 18,000 tons of coal, and more massive facilities of even 100,000 tons do exist.

Although many times derided as simply a "big pile of rock," coal storage facilities are governed by a surprisingly complex set of operational rules. For coal that is to be stored for an extended period of time, layers of coal are "sealed" by interlaying layers of very fine coal to reduce the chance for air and moisture to enter the pile, which otherwise could catch fire or significantly degrade in quality. For example, the most common style of coal storage is simply an open conical pile with the collection point underneath the center of the pile, which can result in a large amount of coal in *dead-storage* between the edge and collection hole. Coal in the dead-storage area does not turn over as rapidly as the coal in the middle, meaning that it has a greater chance to decline in heating value and, thus, usability. To combat this eventuality, the storage facility can periodically move the coal to the collection point with a bulldozer, fill the dead-storage areas with earth-fill (significantly lowering the facility's capacity), or create many more collection points throughout the pile.

Market uses and economics

As evidenced by the sheer physical size of the stocks involved, the historic development coal storage facilities has been more of a physical hedge against supply disruptions than an economic hedge against volatile power markets. In the producing regions, mine mouth coal stockpiles were needed to balance the slower production rates and the ability to fill the unit trains in a timely manner. Along the transportation path, some coal is held at a trans-shipment point for converting the method of travel (barge to rail, etc.) or to produce a blended coal product from one or more varieties. However, the vast majority of coal storage is held at power facilities, where on-site storage prevents supply disruptions from interfering with generation schedules.

Unfortunately, holding a massive amount of coal on-site entails carrying charges—expenses modern power facilities would like to avoid as much as possible. As deliverability of coal becomes more secure, coal supplies at power facilities have slowly declined for many years, with the average coal-fired plant in 2004 having only 40 to 45 days of supply on hand, falling from an average of 60 days on hand during much of the 1990s. Even this is lower than levels common in earlier times of more strident miner unrest and poor transportation reliability, where many power facilities targeted a 90-day (or more) stock level—a period that some say just happened to equal the average length of a United Mine Workers' strike in West Virginia and Kentucky coal mines.

Besides the desire to reduce carrying costs, a more integrated power market has made stockpile management a far more important issue in overall power plant strategies over the last few years—requiring a closer monitoring of stock levels, the greater use of blending coals, and a real-time monitoring of coal quality. One issue in particular has been the focus of this effort—trimming losses because of weathering of coal to provide sufficient availability of high heat-content coal during peak demand periods. This issue becomes more problematic as the seasonal high and low stock levels are taken into account. Seasonal stock levels can fluctuate by one-third during the year; however, some plants can experience far greater variability. This problem became acute during the tumultuous consolidation of the rail industry in the mid-1990s, when a number of power facilities were required to reduce their power output because of low stock levels when deliveries to the plants were delayed significantly—resulting in significant lost sales. As the integration of the power market continues, the operation of coal stockpiles will be increasingly affected by volatile prices in power markets, driving a continued focus on optimizing the stock level as carrying charges vie for availability.

Natural Gas Market

Natural gas has become the most economically important energy resource for the U.S. power industry. Not in sheer volume—coal still reigns by far—but for pricing issues, most power plants on the margin are fueled by natural gas, making it the fuel that sets the price for electricity much of the time. Demand has also grown significantly, with natural gas use growing 250% in the power industry in the past 25 years—80% of the total natural gas demand increase during that period.[3] This demand growth has only been possible through the expansion of the natural gas storage market; expected continued growth of natural gas power generation will fuel the need for even more storage.

The natural gas industry and the need for storage

Wholesale natural gas market. Natural gas is a widely used fuel that provides a quarter of U.S. energy needs. In 2002, according to the DOE, 22.5 Tcf of natural gas were consumed domestically, representing more than a $90-billion retail market. Still dominated by space conditioning demands in both the residential and commercial markets (35% of demand), industrial users hoping to reduce emissions and power generators looking to build combustion turbine–based power facilities have also been drawn by the fuel's promise. The natural gas power generation market has seen the greatest demand growth over the last decade and is expected to continue growing rapidly, with 90% or more of new power plant construction natural gas-based. Currently, power generators use 22% of the natural gas in the United States (up from 17% in 1990), with this ratio expected to expand to 29% by 2020.[4] Space conditioning will likely continue to be a major demand market for natural gas. However, many industry pundits expect overall industrial use to remain stable as industries that use natural gas as fuel stocks, (or other price-sensitive industries) curtail their domestic use of natural gas in response to the elevated wholesale price, which is now, and is expected to remain, far above historical norms.

North America is largely self-sufficient for natural gas use; only 15% of natural gas used in the United States is imported, with roughly 90% of this coming from Canada through pipeline and the remainder entering the market through the four liquefied natural gas (LNG) facilities—through which a growing portion of the new demand is expected to be met (along with new LNG import facilities). As much more imported LNG becomes essential, associated storage at the terminals will also become important. Most natural gas used in the United States currently is produced along the Gulf Coast and in the western states. Although some of it is co-located with oil deposits, the majority is retrieved from natural gas–only deposits in either salt or permeable rock layers. Because the largest demand markets are in the upper Midwest and along the eastern U.S. seaboard, long-distance interstate pipelines are required to deliver the natural gas to market.

With more natural gas being delivered to a wider variety of end users, a marketing and trading component grew up alongside the physical delivery of natural gas. Besides promoting lower costs, reducing volatility, and providing additional choices for consumers, this trading also supported the development of a forward price curve for the market. Initially, the trading was essentially physically based; however, the need and desire for hedging/speculation promoted the development of financial trading that dominates the market today. Because of the integrated nature of the trading, each unit of delivered natural gas on average is actually traded 3 times or more. The financial trading market is far more active, with the number of contracts being traded actually equaling roughly 10 times the amount of natural gas delivered. However, almost 98% of these contracts are settled prior to delivery.

The need for storage. Natural gas storage is the primary means of managing fluctuations in supply and demand and is an essential component of a low-cost, efficient, and reliable interstate transmission and distribution network. Natural gas storage is used for primarily two purposes: to meet seasonal demands for natural gas and to meet short-term peaks in demand that can range from a few hours to a few days. Because there is a distinct seasonal variation to natural gas demand stemming from the heavy space heating demand, storage's impact is significant, making these facilities an integral part of the natural gas industry. For instance, the DOE estimates that more than 15% of all natural gas delivered resides in a storage facility at one time, and on some peak days, stored natural gas delivery accounts for more than 30% of all daily natural gas demand.[5]

Seasonal variation in demand poses a persistent challenge to the market. How does one economically serve a market a few thousand miles away when most of the demand only happens during a few months of the year? Early on, the natural gas industry decided to solve its annual cyclical demand problem by injecting natural gas into underground reservoirs near the demand regions during the summer and releasing it during the winter to meet the peak demands. The first underground natural gas storage field in the United States began operation in 1916 near Buffalo, New York. Providing additional supply during peak demand from natural gas produced in the summer was very attractive to distributors and utilities in the Northeast. At that time, the steel industry was not capable of producing pipelines of the size required to meet the winter peak demand loads of these regions. The only other option was to lay additional, smaller-diameter lines, which would have been uneconomical because they would have been mostly unused during the summer.

Therefore, underground natural gas storage facilities in the downstream market areas were the only feasible option to allow the industry to continue expanding far from the producing regions. As they grew to become an integral part of the supply network, the long-haul pipelines could remain at nearly full utilization throughout the year—which kept the industry profitable—and continued to draw in investment for additional expansion. The DOE, in fact, estimates that natural gas storage is responsible for avoiding 50% of the required transmission system upgrades to the systems.[6] Over the following decades, underground storage capacity continued to expand and was used by pipelines and local distribution companies (LDCs) alike to optimize long-haul pipeline capacity by balancing the flow of natural gas with the injection or removal of natural gas as needed. As the physical natural gas demand increased, so too did the need for more natural gas storage facilities.

Regulatory impact on storage. Regulatory changes have significantly impacted natural gas storage. Before deregulation of the natural gas industry 20 years ago, expansion of the natural gas storage market was used primarily to defer pipeline capacity additions or upgrades; natural gas was simply shipped to and stored in underground reservoirs during the summer for use during the winter months. As part of the restructuring of this industry, natural gas storage was separated from the bundle of services previously rolled into the transmission cost of natural gas, and end users were forced to choose how, or whether, they would use the service. The results have been impressive because as the market changed, so too did storage's role. Storage originally had little capability to respond dynamically to the market. However, in response to the need to support shorter-term storage demands, far more flexible storage facilities have dominated new development.

The origin of today's natural gas industry traces its roots to the Natural Gas Policy Act of 1978, which reversed an earlier decision to establish price controls over interstate natural gas prices, giving the Federal Energy Regulatory Commission (FERC) jurisdiction over interstate pipelines and, in subsequent FERC orders, their storage facilities. In the early 1980s, FERC attempted to instill competition in the wholesale market through FERC Orders 436 and 500, which allowed consumers to contract with the pipeline to reserve capacity for their own use and allowed natural gas producers to sell directly to consumers. By the late 1980s, the federal government forced pipeline companies to break apart their total delivery charge of natural gas into separate services and made plans for a more far-reaching change. In 1992, FERC moved to change the roles of the existing players in the market and created new opportunities and responsibilities for a host of others through FERC Order 636. This later order brought together the open access concepts found in FERC Orders 434 and 500 and outlined the unbundling of services provided by the pipelines by identifying and separating those services—one of those being storage. In Order 636, FERC concluded that natural gas pipeline companies needed to separate their natural gas sales operations from their transportation services, open their trading and transportation information systems to third parties without discrimination, and adhere to new industry-wide standards of behavior aimed at leveling the playing field. With these changes in place, the DOE estimated that 80% to 90% of all working natural gas in underground storage would become available to end-use customers. This had become a primary goal of the federal government in its attempt to foster a secondary market for storage capacity, and, hopefully, additional services to the end-use customer.

The availability of storage to the end user was a key goal of the government because of the optionality storage brought to the market. FERC Order 636 also changed the role of existing participants in the underground storage market and opened the door for a host of new players. Interstate pipelines were allowed to retain ownership and use their storage facilities for load balancing and related services, but since then, these facilities have been required to operate as service providers to any shipper or end-use company and to allow the customer to contract for its own natural gas service. Intrastate pipelines and producers continued to use storage to increase peak demand deliverability and to improve the system reliability because they were regulated by state, not federal, regulations. In both cases, these pipeline firms continued to invest in storage to both enhance the efficiency of their transmission network, and to extend their offerings of higher-margined services. As planned, consumers such as LDCs and large commercial and industrial (C&I) firms began to use storage to supplement their winter season. Through using storage, they could effectively hedge against decreases in natural gas supplies, arbitraging the difference between winter and summer prices with storage fees. Marketers also used storage to support arbitrage and hedging of natural gas contracts; although possessing greater risk, this strategy held out the greatest promise for profit and was, therefore, better suited for independent players than the more regulated actors in the market.

Besides providing the opportunity for additional storage use, FERC Order 636 also changed the economics of natural gas delivery by providing additional incentives for customers to take advantage of storage. This extremely important change raised the cost of pipeline transportation for consumers and resellers with a low load factor (but wide variations in natural gas use) by mandating a straight-fixed-variable (SFV) rate design. SFV essentially shifted all transmission fixed costs to a monthly demand charge for reserving the pipeline capacity. In response, some shippers increased their storage capacity to offset reductions in pipeline capacity reservations. Another impetus for shippers to use storage facilities under FERC Order 636 was that pipeline companies were given significant latitude in penalizing shippers with out-of-balance accounts (monthly true-ups). Shippers with fluctuating demand were again the hardest hit, further promoting the use of short-term storage to maintain balance (load management) and thus avoid penalty.

Besides federal action, state-level unbundling of natural gas pipeline services has also affected the evolution of natural gas storage usage. As state public utility commissions (PUCs) unbundled retail natural gas services, LDCs and other large end-use consumers became more peak sensitive and required more balancing of services from storage facilities for their trading activity across the various shippers.

Facility types and operations

Storing natural gas underground can be accomplished in more than 75% of the United States, with different geological formations supporting different types of natural gas storage facilities:

- **Depleted reservoirs.** Depleted natural gas or oil fields can become storage facilities by simply refilling the permeable rock structure that held previous hydrocarbon deposits, taking advantage of existing wells, gathering systems, and pipeline connections and saving significant start-up costs. Because of their geological stability, there are few leaks during operation.

- **Aquifers.** Some aquifers (mostly in the midwestern United States) have been converted to storage reservoirs. With an impermeable layer above, these facilities create a *bubble* of natural gas within the water-bearing layer. Unfortunately, there is a significantly greater risk of leakage because the original structure did not hold natural gas under pressure. Also, significant amounts of water can be included in the natural gas during withdrawal, requiring additional de-watering equipment.

- **Salt caverns.** The most recent type of natural gas storage facility is an open cavern in a salt dome created by hollowing out the volume with a water jet. Because of the reliance on salt domes, these facilities are limited primarily to areas with suitable salt-bearing geology, found primarily along the U.S. Gulf Coast. A few salt caverns have also been constructed in salt formations in northeastern, midwestern, and southwestern states. Cavern construction is more costly than depleted fields, but the facility has a far greater operational flexibility and capability and essentially no leakage.

The different physical properties of varied geological formations (porosity, permeability, and retention capability) determine how that natural gas will be stored, and at what rate and to what extent the natural gas can be retrieved—leading directly to operational uses for the natural gas in storage:

- **Base gas.** The volume of natural gas that becomes permanent inventory to maintain adequate pressure and deliverability rates throughout the withdrawal season. Some of this natural gas is permanently unrecoverable, trapped by physical forces in the pores of the facility walls.

- **Working gas.** The volume of natural gas that is intended for cyclical injection and withdrawal for sale to the marketplace. Therefore, this is the real measure of interest to the market.

- **Deliverability.** A measure of the amount of natural gas that can be withdrawn from a storage facility daily. The deliverability of a given storage facility is variable and depends on factors such as the pressure within the reservoir, compression capability of the reservoir, and so forth. In general, a facility's deliverability rate varies directly with the total amount of natural gas in the reservoir; it is at its highest when the reservoir is full and declines steadily as working natural gas is withdrawn.

Market uses and economics

Storage facilities located in different geographical areas lend themselves to different market roles. Like oil storage facilities, natural gas storage facilities in production regions allow for the field's output to be disconnected from production; otherwise, the output of the wells must be massaged to match demand. This lowers the life of the well and requires fewer, but more productive wells—improving the economics of the field. Along the pipelines and in the distribution markets, storage facilities provide much-needed physical balancing services. This enables pipeline companies to optimize pipeline pressure and overall performance by maintaining as constant and high a flow rate (pipeline utilization) as possible—allowing pipeline firms to maintain service with lower capital investments.

Marketing diverse facility types is also tied to their varied capital and operating costs. Because the majority of natural gas storage is used for seasonal heating demands, depleted natural gas reservoirs and aquifers make up the largest component of underground natural gas storage, comprising 95% of the total working gas. In 2000, it cost on average $0.48/MMBtu per season to store natural gas in a depleted field.[7] These facilities cycle once per year, injecting natural gas from May to October (214 days) and withdrawing it from November to April (151 days). Because it takes approximately 180 days to fill and approximately 120 days to withdraw, the operators and customers of these facilities are generally price insensitive as they have very little operational flexibility. With the capability of only one cycle per season, the profitability of the facility becomes more of an issue of seasonal price changes than operational activity. Because of the low deliverability of these primary storage facilities, the amount of working gas in storage has become a leading indicator of short-term natural gas prices, especially during the later half of a heating season when stock levels—and thus deliverability—make the market more susceptible to volatility in a cold snap.

Unfortunately, the increasing demand from natural gas–fired power facilities (mostly during the summer) is putting stress on this historical pattern and affecting the seasonal demand injection/withdrawal schedule. As more natural gas–fired plants are brought online to supply power for cooling loads, these plants compete for natural gas supplies in the middle of the injection season. The need for more flexibility—higher deliverability—has, therefore, put a premium on salt-dome storage facilities. These facilities only require approximately 20 days to fill and approximately 10 days to fully withdraw the natural gas in storage (depending on size). Although they represented only

4% of total working gas capacity in 2002, they provided 17% of total deliverability.[8] With their faster deliverability, the salt-dome facilities provide a host of load management services:

- Peak supplies for sudden weather shifts

- Load balancing for shippers (to avoid paying imbalance fees)

- Emergency supply backup in case of supply disruption (hedge for end users against price changes)

- A tool for marketers to speculate on price changes

Although salt domes have high capital costs, their greater flexibility gives them a much lower operational cost—making them profitable over a wider range of market conditions, especially multiple cycling throughout the season. For instance, salt-dome storage costs $1.08/MMBtu for one cycle per year, but this cost drops to $0.28/MMBtu when the facility is cycled five times.[9] Therefore, this ability to cycle its working gas several times in a season can easily offset the additional cost of high-deliverability storage—if the operators correctly anticipate most of the market price swings. With this flexibility from high-deliverability storage, the value potential of the natural gas in salt caverns changes from simple long-term arbitrage to market timing—something of far more value.

Lessons Learned

Groups involved with setting public policy and developing energy storage facilities in the power industry can learn much from how other energy markets have integrated storage into their value chain. These storage assets have proved to be a key component for providing reliable service, low prices, and flexibility in each of these other energy markets. Although the power industry is obviously quite different in many ways, many of the economic forces are similar (especially in the natural gas industry), leading to operational insights and leverageable market strategies.

In these other energy markets, the introduction and use of storage has undergone a surprisingly similar learning curve. Although initially used to level variations in supply, a more integrated use of these facilities has resulted in fewer required transmission upgrades and lower system expansion costs. Deregulation in these markets—again with particular respect to natural gas—enabled the storage facilities to provide ready reserves and create greater customer choice, and, surprisingly, it created a greater demand for the use of storage facilities. As wholesale physical trading begat even more financial trading, storage assets served not only as the basis for many of these contracts, but also as a tacit physical backstop for the trading activity.

Although the petroleum and coal storage industries provide valuable lessons, the natural gas industry has especially valuable and relevant lessons to teach the electric power industry in the use of storage in the wholesale and transmission markets. There, the use of storage facilities first was recognized as a means to lower network expansion costs, and they have been incorporated into a nationally integrated production, transmission, and end-use service network. This did not happen overnight, but rather through a combination of regulatory prodding and innovative use of new technology. This storage capability has proved to be a crucial part of the natural gas industry—without it, the industry would not currently be able to deliver the capacity or provide the flexible services that were required for market expansion. Nor would the natural gas industry have been able to emerge so successfully from restructuring without a storage component to dampen a wholesale market, provide optionality, and raise the level of confidence in the market. As the electricity market looks to complete its own restructuring (or simply start again), it can look to this history in the natural gas industry for guidance.

The most valuable lessons learned for the electric power industry are focused along three avenues:

1. Value of the storage asset
2. Impact of technological advances
3. Impact of regulatory reform

Through reviewing the experience of these markets' use of storage assets as they underwent change, groups looking to expand the presence of energy storage assets within the electric power industry can gain some significant understandings in what works, and what does not.

Value of storage

The value of a storage facility is linked to its usefulness, which is related to the

- Speed of the market
- Scale and location of the facility
- Deliverability of the facility

These criteria dictate much of how a storage facility is able take advantage of changing demand patterns and market activity to provide enhanced services to the customer. As a market matures and integrates storage deeper into its operations, the uses, as well as the values derived from those uses, will change. The maturity of the industry's use of storage is thus implied by the leveragability of these assets.

The speed of a market describes the rate at which the price and availability of a commodity change. Slower, physical-delivery commodity markets can largely use the *flow* of the commodity only to replenish physical demand or rebalance a shortfall. However, as activity in the market increases, the ability to leverage the stored quantities of the commodity increases and raises a storage facility's inherent value. For instance, a highly valued storage facility could take the form of a highly deliverable ready reserve to rebalance a supply-demand imbalance—one where it would provide a value through action or as a reserve unit. This speed-to-value index is taken to the extreme example in the electric power industry, as there is effectively no commodity flow to replenish a shortfall; demand cycles that happen daily here can take weeks or months in the natural gas market.

The storage facility's location and scale are other aspects that affect how the unit will be used, with increased use translating into higher value. One of the initial *values* ascribed to a new storage facility is its ability to preclude the outlay of additional capital for building or refurbishing transmission facilities that would be used to supply peak demand periods—because the peak demand occurs infrequently, the entire system is then vastly underused. In the natural gas industry, storage has been rightly credited by the DOE with avoiding 50% of the required transmission system upgrades and allowing the average utilization of the system to remain more than 90%.[10] This contrasts with past electric utility strategies that dictated generation and transmission capacities to be so overbuilt that the system's current utilization rate is effectively capped at 60%. Positioning a cache of supply past a capacity-constrained portion of the transmission system also creates additional value to the commodity if it can be quickly delivered. This strategy of capturing locational scarcity rent was a central theme in many merchant power facilities that were built during the 1990s construction boom looking for transmission cul-de-sacs or *load pockets*.

Finally, the last issue driving the value of a storage facility is the deliverability of that unit. For some applications, a slow withdrawal capability means the facility is essentially useless, leading to a poor value attributed to that asset for that particular application. With greater deliverability, the range of applications grows, making the unit more flexible and responsive to market needs. For instance, experience with salt-cavern storage facilities has also shown an increased number of shorter-duration injection and withdrawal activities throughout the year, expanding the possibilities for new roles as a swing-supply—especially for the growing demand from power facilities. As high-deliverability storage has become integrated into short-term supply strategies, the market has shown its increasing comfort level in storage's deliverability by holding lower storage inventories. Figure 2–1 shows that as storage throughout the 1990s grew to provide first 10% and then 15% of all demand—increasingly from a growing number of high-deliverability salt-dome facilities—the amount of natural gas in storage actually fell as a portion of total use, signifying the stabilizing component storage provided to the market.

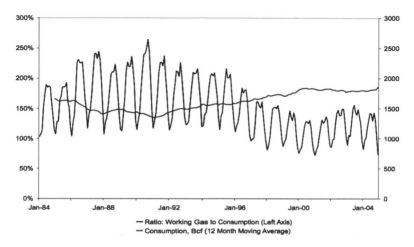

Fig. 2–1. High deliverability enhances the value of storage (Data: US Department of Energy).

Technological change

Technological change can be driven both from inside and outside a particular market. When the natural gas industry was new, the inability of the steel industry to produce pipes large enough to meet summer demand levels drove the need to store large quantities of natural gas underground near the demand regions. This quandary led to a much higher use and efficiency of the entire industry. Subsequently, ever-larger coal facilities in the 1950s and a declining competitiveness of coal in general because of high transportation costs led to the unit train delivery of coal—requiring much larger and better managed coal stockpiles. Because of the improved reliability of the entire supply chain, coal stockpiles have been able to decline for more than 20 years, greatly lowering the carrying cost of the coal and thus improving the operating margins of the facilities.

New technologies can change the existing economics in the market, which in turn make the introduction of more innovative technologies easier. The introduction of underground natural gas salt-dome storage facilities is a good example of how new storage technology with a lower operating cost can alter the use of storage within a market. One of the central benefits of salt-dome storage facilities is the much higher deliverability over other types of natural gas storage—allowing these facilities to provide multiple injection and withdrawal cycles per year. Although these salt-dome storage facilities are more capital intensive, their ability to have upwards of 10 cycles per year allows for the average injection and withdrawal cost to be lower than the depleted natural gas field storage facilities. For instance, according to Simmons & Company International in 2000, it cost on average $0.48/MMBtu per season to store natural gas in a depleted field. That same cost was $1.08/MMBtu in a salt-dome facility for 1 cycle per year, but this dropped to $0.28/MMBtu when the facility was able to cycle 5 times per year.[11] This advantage allowed salt-dome facilities to spring from essentially 0% of the market in 1990 to more than 15% of all natural gas in storage by 2002, and nearly 20% of deliverability capability for all storage facilities.

The electric power industry is arguably the energy market most affected by technological change, with this change coming through either revolutionary (disruptive) or evolutionary means. However, as the production and delivery of power is an infrastructure-oriented market, the electric power industry is hardwired toward accepting evolutionary, rather than disruptive technology. Because electricity storage technologies are more geared toward improving the efficiency and operational capability of other components in the system rather than "establishing a new paradigm" in how the industry operates, these energy storage technologies should find a receptive audience, providing the technology is mature and cost-effective.

Regulatory change

Regulatory change can have a profound effect on the use of storage facilities. Because of the importance of energy markets to the national economy, regulators have long looked for ways to leverage any assets within a particular market to create greater customer choice and lower costs. For instance, in the petroleum industry, strategic petroleum reserves have been created (in many countries) to act as a buffer for major supply disruptions. Acting together, the groups controlling these reserves can have a large impact on the market, which makes their mere presence a means to lower supply uncertainty and, hence, prices. Regulatory changes are at the heart of the wholesale power and transmission market opportunities facing energy storage in the electric power industry today.

Regulatory action within the natural gas industry provides a good example of how policy makers can successfully leverage storage assets. Consumer demand for contract relief and for more choices in natural gas delivery options led the federal government to desire a more self-regulated and independent storage market. By forcing the unbundling of the delivery charges (FERC Order 636), the transportation, storage, handling, and delivery services were then priced competitively; natural gas storage facilities were able to operate as third-party service providers in a deregulated wholesale market, serving a much larger number of end users. After delivery charges were unbundled, demand for more reactive storage services expanded. It is widely believed that the natural gas industry would not have been able to emerge so successfully from its restructuring without a storage component to dampen a potentially volatile wholesale market, provide optionality for sellers and buyers alike, and raise the level of confidence in the market.

With the passage of FERC Orders 888, 889, and 2000, the electric power industry finds itself in a position similar to that of natural gas after FERC Order 636 was passed. Unfortunately, storage assets simply do not exist in the electric power industry to the same degree they did in the natural gas industry prior to deregulation. Because of the inability to economically transmit electric power more than a few hundred miles (normally), the electric power industry developed into a far more vertically oriented industry compared to the horizontally oriented natural gas industry—leading to the early recognition of the value of storage facilities by a wider number of market participants in the natural gas versus the electric power industry. For this reason, a single natural gas storage facility was able to serve a number of customers easily once the delivery charge was unbundled.

As the electric power industry moves toward a similar horizontal focus, large-scale energy storage facility development may mimic the experience of natural gas storage facilities. For example, current pumped-hydro facilities provide significant power grid stabilization services to the utilities that own them, but if they were slated to operate in a merchant role through an ISO, their stabilizing services could be provided to a far wider audience. For the power industry to mimic the natural gas industry, ancillary services must be further broken out—in a fuller and more systematic way (i.e., replicated nationally)—from the delivery charges. With the market determining the cost of (and need for) these essential services, this action would significantly support the development of a self-regulating market. As in the case of early natural gas storage, federal involvement will be key; not to give electricity storage facilities an advantage, but simply to level the playing field to create a market for their services.

References

1. Energy Information Agency. 2002. *Annual energy review 2002*, 159–211. Washington, DC: U.S. Department of Energy.

2. Energy Information Agency. 1997. Causes and effects of lower inventories. *Petroleum 1996: Issues & trends*. Washington, DC: U.S. Department of Energy.

3. Energy Information Agency. 2004. Natural gas. *Annual energy review 2002*, 213–232. Washington, DC: U.S. Department of Energy.

4. Energy Information Agency. 2004. *Annual energy outlook 2004*. Washington, DC: U.S. Department of Energy.

5. U.S. Department of Energy. *Gas storage white paper* (draft). Washington, DC: 6100 Fossil Energy Working Group.

6. Ibid.

7. Dietert, J. A., and D. A. Pursell. 2000. *Underground natural gas storage*. Houston, TX: Simmons & Company International.

8. Energy Information Agency. *Annual energy review 2002*, 213–232.

9. Dietert and Pursell. *Underground natural gas storage*.

10. U.S. Department of Energy. *Gas storage white paper*.

11. Dietert and Pursell. *Underground natural gas storage*.

3 — ELECTRICITY STORAGE TECHNOLOGIES

As the name implies, energy storage technologies absorb electrical power during periods of excess capacity so that it can be released later when it has more value. When released, the energy can either be delivered in large amounts for commodity use, or in a measured, controlled manner to optimize the operation or enhance the reliability of the power delivery system. The storage unit can also be run by constant charge/discharge cycling to dampen unwanted fluctuations in the system's power level. Alternating current electricity itself is not stored directly, but rather, it is converted and stored by mechanical, chemical, or electrical potential energy methods. Each of these storage methods has its own particular operational range and capabilities, and the use of a particular method usually predisposes the storage unit into a set of applications for which the facility's technology is best suited.

System Components

Each energy storage unit facility is made up of three distinct components: the storage medium (or subsystem), the power conversion system, and the balance of the plant. The size and cost of each of these components vary, based on the setting and the application for which it is used, leading to a range in costs even within one technology. These costs will decline as the core storage technology matures, improvements are made in the production processes, and economies of scale such as production levels continue to rise for many of these recently commercialized technologies. In the near term for technologies in pilot-phase installations, initial engineering and installation costs can also be significant because of the lack of experience—a cost component that also declines with wider commercialization.

Storage medium

The heart of every energy storage facility is the *energy reservoir* or storage medium, which can take the form of mechanical, chemical, or electrical potential energy. The differences in how energy is stored in the various storage media help to define each technology's capabilities, leading to one or more technologies being better suited for certain applications than the others. For instance, large-scale storage technologies that can arbitrage energy between long time periods rely on air or water because they both are low-cost and both can be stored a long time without loss. Storage media that require greater energy expenditures for support systems such as necessary environmental controls for the storage medium would not be suitable for such roles. However, if their storage medium was able to cycle through a charge/discharge cycle quickly and repeatedly without damage to the unit, they would be a good choice for other applications.

Storage medium costs vary widely because the different media vary significantly from air to chemical electrolytes. An important aspect to their costs is the energy density of the medium; high-density storage media allow for smaller supporting equipment, whereas lower-density material requires a large storage facility, with its associated expenses. Structurally, the costs for each storage medium can generally be broken down into two components—the initial capital cost of the medium itself (including replacement material), and the costs associated with maintaining the storage medium in a charged state until a discharge is requested. For most systems, the cost of the storage medium is one-half the cost of the entire facility, and it is expected to remain in that range for the foreseeable future. The reason for this is that as the level of technological progress continues to improve, the cost of the overall unit will decline (other subsystems are improving as well) for the same application, or more capable units are developed to address larger applications.

Power conversion system

All storage technologies, except for some mechanical energy storage systems (i.e., pumped-hydro storage or compressed air energy storage [CAES]), need dedicated power electronics for alternating current–direct current (AC–DC) and DC to AC conversion (in pumped-hydro and CAES plants, these are part of the motor/power generator train). A power conversion system (PCS) acts as the electrical interface between the customer or utility power grid and the storage subsystem, and consists of such components as a converter, DC and AC switch gear, and programmable high-speed controllers.

The PCS of an energy storage subsystem has two major roles. First, the PCS is used to convert the AC electrical energy of the power grid into DC for storage. When the storage system is being charged, the converter acts as a rectifier (changing AC to DC); during discharge, the process reverses and the converter operates as an inverter, changing DC to AC. Second, the PCS conditions the power during conversion so that no damage is done to the storage facility or the customer's electrical system. A related system of power electronics monitors and controls the condition of the storage subsystem (this is of a greater concern when the storage medium is a chemical-based battery that can be damaged through misuse). The actual PCS for any one particular storage technology or installation is heavily dependent on the balance of capabilities desired (reliability, responsiveness, etc.), which, in turn, affects the cost of the unit.

The U.S. Department of Energy's (DOE) storage program has identified the customization of the PCS for each storage technology as one of the most important energy storage system integration issues. Because each energy storage technology operates differently during charging, standing by (floating), or discharging, the life span of the storage facility will be greatly reduced if the PCS does not take that into account. In particular, the design and cost of a PCS greatly depends on the primary activity of the storage facility, and thus most PCS components are generally compared on a dollars-per-MWh (commodity arbitrage) or dollars-per-MVA (power quality) basis, depending on the need for real or reactive power (which the PCS provides).

In terms of capital expenses, PCS costs for standard-scale installations are typically in the range of $100 per kW to $200 per kW for those focused on energy arbitrage (MWh) to $200 per kW to $400 per kW for those focused on power quality (MVA). Because customization is still the rule rather than the exception, the PCS often represents anywhere from 33% to nearly 50% of the entire cost of the storage installation (depending on the application and scale of the facility), but continued effort keeps reducing the costs of these units. With advances toward modularization and mass production of these units, unit costs are expected to decline further, with those systems focused on energy management estimated to see declines of 10% from current levels. Power-quality PCS systems could see price declines in the coming years of up to 25%.

Balance of plant

The balance of plant (BOP) encompasses the facility (and control systems) to house the equipment, the environmental controls, and the electrical connectors between the PCS and the power grid. This aspect of an energy storage installation can vary tremendously based on the requirements of not only the technology, but also the application. For example, the BOP typically includes transformers, electrical interconnections, surge protection devices, a support rack for the storage medium, the facility shelter (or component of total), and environmental control systems. Because storage media can operate throughout a wide range of environmental conditions, the space conditioning is sometimes the target of cost cutting. Although undersizing of the electrical components is not often an option, the supporting, particularly the environmental, controls in existing (normally lead-acid batteries) energy storage installations are sometimes lacking, especially if the storage unit is operating in a reserve or backup power mode. Unfortunately, savings here will many times shorten the operating life and hence increase replacement costs, driving up the overall cost for the use of the unit in question.

For these reasons, the BOP is by far the most variable cost component of an energy storage facility. In general, these costs tend to rise when the installation is more customized and are less when modular storage units are used. Normally, however, the BOP costs represent 10% to 25% of the total cost for a typical storage facility. For example, lead-acid batteries should be maintained within a certain operating range to prolong their operating life, necessitating space conditioning for the sometimes-significant floor space required to house the units. Alternatively, flywheels are far more compact and can operate in a much wider temperature range—mostly obviating the need for space conditioning of the unit.

Engineering, procurement, and construction

Similar to the BOP costs, the engineering, procurement, and construction (EPC) costs to build an energy storage facility are highly dependent on the complexity of the equipment and the amount of preparation required for the site. Overall, these engineering and construction costs can add an additional 10% to 15% (or more as the uniqueness level rises) to the total cost of the facility. Some reduction in these costs will be expected over time as engineering service companies gain experience installing energy storage technologies, and the continued modularization of the PCS and storage subsystems allows for a *plug-and-play* installation.

Pumped-Hydroelectric Storage (PHS)

Summary

Pumped-hydroelectric (hydro) storage (PHS) is the oldest and largest of all of the commercially available energy storage technologies. This technology also retains the largest installed base (by capacity) of any storage technology with more than 20 GW in the United States, where it currently represents roughly 2.5% of total summer generating capability. Conventional PHS facilities (fig. 3–1) consist of two large reservoirs, one located at a low level and the other situated at a higher elevation. During off-peak hours, water is pumped from the lower to the upper reservoir where it is stored. During peak hours, the water is released back to the lower reservoir, passing through hydraulic turbines to generate electrical power. These facilities generally operate on a daily schedule, with some facilities also operating as a conventional hydropower facility for irrigation or other public uses. Older designs had round-trip efficiencies of more than 60%, but repowering of some plants recently has pushed this higher than 75%. Although it is a proven and valued technology, difficulties in siting new facilities and total system costs that can reach upward of $2,000 per kW curtail most future prospects for development, especially in developed countries.

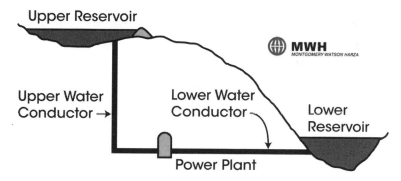

Fig. 3–1. Pumped-hydro storage (Courtesy of Montgomery Watson Harza (MWH)).

Historical origins

PHS units first appeared in Italy and Switzerland in the 1890s, and arrived in North America in 1929, when the 32-MW Rocky River pumped-hydro storage facility in Connecticut entered service. Moving into the 1950s, PHS facilities were becoming more widespread as the designs for them began incorporating a single reversible pump-turbine— previous designs incorporated separate pump impellers and turbine runners. Besides being more capital intensive, these previous designs limited the unit's reaction time to reverse direction as the channel not being used had to be drained of water. The peak construction period for these facilities in North America stretched from the 1960s through the 1980s, as utilities found these units to be a valuable way to manage system loads throughout their service territories. Pumped-hydro facilities were, in fact, the only commercially available energy storage technology for utility-scale applications until the 1970s. Interest in pumped storage waned in the 1990s because of the technology's high capital costs and difficulty in locating new operating sites because of environmental concerns.

Design and operations

A typical PHS facility consists of two large reservoirs, one located anywhere from 30 m to 650 m above the base level, with 300 m of hydraulic head (the difference between the two reservoirs) generally considered the preferred height for new development. These facilities can either use freshwater or seawater as the working medium. As PHS facilities share a large portion of their equipment with the much larger conventional hydroelectric industry, this has allowed them to take advantage of advancements in variable-speed motors and improved impellers to increase their capability and efficiency.

During off-peak hours, water is pumped from the lower to the upper reservoir, where it is stored. To generate electricity during peak demand hours, the water is released back down to the lower reservoir, passing through hydraulic turbines and generating electrical power. Energy in PHS facilities is stored as potential (mechanical) energy; water is the working fluid, and gravity provides the driving force. Whereas the generating capacity (kW) is a function of the rate of water outflow and the hydraulic head, the energy (kWh) that can be stored in the facility is a function of the reservoir volume and hydraulic head. Therefore, attaining the greatest hydraulic head for the facility is normally the most critical design issue, because the ability to expand the upper reservoir (to expand to the facility's storage capacity) is usually more expensive and difficult. This aspect allows variations in the design for a facility with the same storage capability; a larger head requires a smaller volume of water to provide the same amount of energy storage as a larger and lower reservoir. Increasing the hydraulic head also requires smaller-diameter turbines and the accompanying capital equipment. Assuming that a wide, straight pipe links the two reservoirs (to reduce frictional losses) and that evaporation is minimal, the main losses in PHS storage operation stem from the inefficiencies of the pump and turbine.

PHS facilities possess a fast reaction to go along with their large size, providing a unique capability to the system operators, which allows them to cover a wide range of applications within the wholesale power market—including providing reliable system-wide frequency regulation and contingency reserves in addition to arbitraging commodity power for sale. Although the water held in the upper reservoir could be held there indefinitely, PHS facilities generally operate on a daily schedule—although it is not uncommon for a facility to operate upward of eight times per day. Compared to generation units, these facilities are able to quickly respond to changing market conditions, as short as 10 minutes or less from complete shutdown (or full reversal of operation) to full power; maintained on standby, these facilities can even reach full power within 10 to 30 seconds. As these facilities can also reverse the direction of flow quickly, they provide an important balancing role for system stability as well, switching from absorbing a few hundred MWs of excess system power to producing an equal amount quickly in case of an emergency. Over a longer-term horizon, this strategy is also used to balance or optimize thermal facilities elsewhere on the power grid. These attributes help PHS facilities maintain an availability rating that can surpass 98%, and a forced outage rate below 1.5%.

Whereas older designs had round-trip efficiencies of less than 60%, improved designs coupled with repowering with modern control equipment have recently pushed this higher than 75%. Besides losses incurred as the water travels through the motor/turbines, the efficiency of these facilities is also greatly affected by the effective head loss because of friction and turbulence in the pipeline connecting the upper and lower reservoirs; shorter distances and larger pipes incur less energy loss. The cycle time requirements for these facilities can be expressed as 1:1 for charge/discharge at full power, but operational requirements of the local power grid normally dictate the use of the unit as a flexible system optimizer instead of a rapid commodity energy sink/source. The effective cycle life of these systems is normally viewed in decades of operating rather than a set number of cycles, as uses such as providing frequency regulation would not require full charges/discharges; component replacement and/or upgrades are more along the lines of other power facilities.

Cost issues

PHS facilities are capital intensive, as their costs include not only significant civil construction, but also reversible pump-generating units and reservoir(s) that require frequent drawdowns. In total, these requirements push the average capital costs to $1,500 per kW to $2,000 per kW for development costs. Because most pumped storage facilities are developed with many hundreds of MWs, the total cost for such a facility will normally reach into the hundreds of millions of dollars in developed countries such as the United States or in Europe, where environmental mitigation requirements exist. However, with much of the capital equipment already in place, existing sites provide extremely cost-effective upgrade opportunities of core components such as the impellers. To improve the economics of this technology, developers and engineers are looking at creative ways to reduce costs when constructing new projects. Some of these ideas have included using exhausted mines as the lower reservoir (including sealing the cavern walls) or adding a PHS unit to an existing hydroelectric facility.

Installations

Typical PHS turbines range in size from 30 MW to 350 MW, with facility capabilities ranging from 300 MW to 1,800 MW. Currently more than 90 GW of capacity exist in more than 240 PHS storage facilities around the world—roughly 3% of global generating capacity. This is an increase from 1990, when only 74 GW of installed generating capacity existed. The current 20 GW of capacity in the United States (up from

15 GW in 1990) represent 2.5% of the nation's summertime generating capability. Other developed countries have also invested heavily in these facilities, with Japan maintaining roughly 10% of its total generating capacity in PHS storage facilities, and European Union (EU) countries maintaining more than 32 GW of PHS storage capacity.

Example—Rocky Mountain, GA.[1] The Rocky Mountain facility—developed in 1991—is a three-unit, 848-MW pumped storage hydroelectric plant located in Floyd County near Rome, Georgia. Oglethorpe Power's Rocky Mountain pumped-storage facility was originally intended to run in a peaking mode, but in its first calendar year of service ran every day except one to take advantage of strategic power marketing deals. The facility won the 1998 Powerplant Award from *Power Magazine* for demonstrating the value of large-scale energy storage to a regional power market.

Example—Dinorwig, Wales, UK.[2] The Dinorwig plant in Wales, UK, is one of the most well-known pumped storage plants in the world. It was constructed between 1976 and 1982 in Europe's largest man-made cavern under the hills of North Wales. Each of the station's six generating units acts as both pump and turbine, delivering 317 MW of power, sustaining 1,800 MW for a total of five hours. Working volume for the facility is 6 million cubic meters, with a head of approximately 600 m. If held as spinning reserve, the entire plant can reach maximum output in less than 16 seconds.

Example—Okinawa, Japan.[3] The world's first seawater PHS plant was built in Kunigami Village, Okinawa Prefecture, Japan, in 1999 with a rated capacity of 31.4 MW. Taking advantage of Japan's abundance of coastline, these facilities can be located in rural areas near power facilities to lower power transmission losses. This PHS storage facility was built by the Electric Power Development Company for the Agency for Natural Resources and Energy of the Ministry of International Trade as a pilot facility. The facility's upper reservoir is located 500 m from the ocean on a 150-m cliff. The facility possesses a 160-m working head, and the powerhouse is located halfway between the reservoir and the ocean, 150 m below the surface. Because the working fluid is saltwater, significant corrosion protection was included in the design, such as simplifying the design of the interior and the inclusion of cathodic protections to prevent crevice corrosion.

Prospects and challenges

Prospects for this technology are limited in most developed counties because of high development costs, long lead times, and reservoir design limitations, such as environmental concerns and large land needs requiring placement far from load centers and existing transmission facilities. International prospects for development are better but also are affected by these issues, plus the requirement for a sufficiently developed power grid with a sufficient difference in on- and off-peak prices and demand levels to make development economically viable. Currently, new PHS facilities are continuing to be developed in countries with rapidly expanding power sectors, as in China, where 3 GW of capacity are being added to its existing total of 2.4 GW.

According to Peter Donalek, system studies manager at MWH Global (a developer of PHS facilities), one of the more important recent developments for PHS technology is the introduction of adjustable speed machines for the pump/turbine in both new and retrofit projects. This advancement provides a number of benefits over previous, fixed-speed pump/turbines. Operational benefits include both partial speed pumping (60%–100%) and the ability of the facility to provide frequency regulation in both pumping and generating modes (previously this was just available during generation). This flexibility allows the unit to absorb power in a more cost-effective (and for the power grid, useful) means, and increases not only the pump-turbine life, but also produces less stress on supporting equipment such as the seals and bearings. Overall, using adjustable speed machines can improve the efficiency of a pumped storage facility by 3% or more because the unit can be optimized for peak efficiency in both pumping and generating modes.

One proposed variant of PHS technology design (primarily from Japan) uses an underground cavern instead of a free body of water as the lower reservoir. Because these facilities can be isolated from existing water resources, there is significantly more freedom in the choice of plant sites, and a reduced impact on the environment. These facilities are designed to move the same water up and down in a directly vertical direction, giving a greater energy-per-unit volume than a natural system—which must pump the water up at an angle and is limited by geography to a maximum vertical distance. This also means, however, that the head must be significantly higher to make up for the reduced volume of working fluid.

Although many U.S. utilities have expressed a wistful desire for more of these facilities, the current focus for development of this technology in countries like the United States is to upgrade existing PHS facilities. Examples like Ameren UE's upgrade of the Taum Sauk facility (on the Black River in Missouri) show that utilities can add more than 100 MW of capacity for $250 per kW or less. That these facilities continue to be viewed as premium resources can be attested to by the prices garnered during auctions. Current upgrades for these facilities normally include uprating existing facilities with advanced pumps/turbines, impellers, control systems, and variable-speed drives to increase capacity by 15% to 20% and operating efficiency by 5% to 10%, and to improve the unit's charging rates.[4]

Major developers

Although no firm is actively developing large PHS facilities in the United States, firms such as MWH Global and American Hydro are actively upgrading existing facilities, such as the Taum Sauk plant previously mentioned (American Hydro was retained by Ameren UE to do the upgrade between 1996 and 1999; original construction was in 1963). Replacing the impellers and other work increased turbine capacity (from 350 MW to 450 MW), round-trip efficiency (43% to 70%), and station flexibility at a cost of only $25 million—effectively representing a new 100-MW power plant for only $250 per kW.

Compressed Air Energy Storage (CAES)

Summary

Compressed air energy storage (CAES) systems use off-peak power to pressurize air into an underground reservoir, which is then released during peak daytime hours to power a gas turbine for power production. In a gas turbine, roughly 66% of the power produced is required to compress the air for combustion and high-temperature expansion. The strategy behind this technology is thus to substitute lower-cost energy from an off-peak baseload facility for the more expensive natural gas fuel used to power a separate compressor to precompress and store the air in an underground chamber. The air is later fed directly into the expander combustion train (without a compressor) to produce electricity. This allows a CAES facility (fig. 3–2)—without the parasitic compressor load—to produce essentially three times the electricity as a gas turbine from the same amount of natural gas. To improve the efficiency of the unit, the exhaust gas is passed through a recuperator to preheat the air coming from the high-pressure storage cavern. CAES and PHS facilities are the only storage technologies in commercial operation able to provide large-scale storage deliverability (more than 100 MW) for use in the wholesale power market.

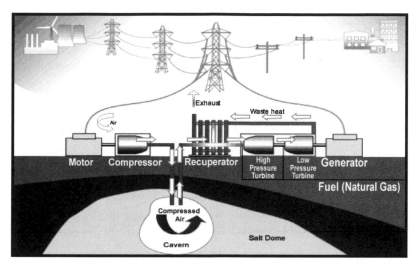

Fig. 3–2. Compressed air energy storage (CAES) (Courtesy of Ridge Energy Storage)

Historical origins

CAES technology development can trace its roots to the early 1960s during the early evaluation of gas turbine technology for power production. The technology gained additional support during the 1970s to provide load-following and peaking-power support because of its promising fuel efficiency and response capabilities compared to early natural gas turbines. Because of its load requirements during off-peak hours, the technology was also considered a means to provide additional off-peak demand for the growing number of nuclear power facilities to improve their overall utilization rates. The first unit (which is still in operation) was developed in Germany in 1978, followed by another unit in the United States, completed in 1991. A number of other potential follow-up projects around the world were investigated, but none came to fruition. However, the continued development of gas turbine technology, especially when expanded into the combined-cycle concept, crowded out much of the subsequent interest until recently.

Design and operations

A CAES facility consists of five major above-ground and underground components:

1. Power train motor/generator that employs dual clutches to provide for the alternate engagement of the compressor and turbine

2. Air compressors, which require low pressure (LP) and high pressure (HP) casings in two or more stages (intercoolers and after-coolers) to achieve efficient compression and cool the air before discharge into the cavern (which also reduces the moisture content of the air)

3. Expansion or expander turbine train (high- and low-pressure turbines)

4. Controls for operating the combustion turbine, compressors, and equipment to handle the switchover from charging to generation

5. Balance-of-plant auxiliary equipment consisting of fuel supply, mechanical, electrical, piping, and cooling systems

A recuperator (part of the power train) air-to-air heat exchanger is used to boost the efficiency of the system by preheating the cavern air before it enters the first expansion stage. The recuperator heats the inlet air with exhaust from the turbine to allow the first expansion to meet the needed inlet temperature and pressure to the combustors. The facility also contains small inlet air heaters to keep the expander and recuperator warm for a fast response to full load. The secondary compression in the cavern as the pressure increases adds temperature, which is offset by cooling in the wellhead control valve on release of the air. The reservoir used by the system can be a salt cavern, an abandoned hard-rock mine, depleted gas fields, or an aquifer—although this last setup would only allow for a low-pressure release of air (550 psi vs. 1,200+ psi for salt cavern or mine) and a subsequent lowering of energy storage capacity of the unit. Man-made underground storage caverns can also be used, having either a rubber or steel layer along with a concrete lining to contain the air.

CAES systems use off-peak power to pressurize air into an underground reservoir, which is then released during peak daytime hours to power an air expander/gas turbine for power production. Normally, roughly 66% of the power produced in a gas turbine is required to compress the air for combustion and high-temperature expansion. Instead of using high-cost natural gas for the turbine to compress the air when the gas turbine is operating, the strategy behind this technology is to use off-peak power from low-cost sources (generally baseload coal and nuclear power facilities, but also including wind power) to run a separate compressor to precompress and then store the air in an underground chamber. This saves the increasingly expensive natural gas fuel to only be used when the compressed air is released from the underground reservoir and fed directly into the expander combustion train—without the parasitic compressor load—to produce electricity. Using the air compressed with off-peak baseload power, a CAES facility can produce roughly three times the electricity of a normal gas turbine from the same amount of natural gas, giving the facility a heat rate of 4,250 Btu per kWh HHV or less during operation. Other operational benefits accrue to the unit because of the high-pressure air reservoir and the off-peak power. These units have a far greater response time and operating range, allowing them to operate with no ambient de-rating and with shorter ramp rates at times—for instance on a hot summer day—than a gas turbine. Because of the lower fuel consumption during power generation of these units, the direct emissions per MWh are substantially lower than the equivalent capacity from gas turbines installed in a peak or mid-range combined cycle plant.

Because of their large scale and fast reaction capability, CAES facilities are among the most flexible sources and sinks for bulk energy in the wholesale power market, making the units able to undertake frequent startups and shutdowns, swinging quickly from generation to compression or simultaneous compression and generation for load optimization. As a source, a CAES facility can provide low-cost energy on the scale of PHS units with an operational range better than a gas turbine. Because the air is prepressurized, there is no derating of the facility because of increasing altitude or summer ambient temperatures (because of less-dense air). This is a major advantage for a peaking facility targeted to operate during the peak electricity periods (summer cooling electrical loads) because for a typical combustion turbine, a 15% derate penalty can occur for a 40°F rise in ambient temperature. A combined cycle facility under the same conditions would suffer a 10% derating. Conversely, CAES facilities do not suffer appreciably from a rising heat rate under partial load as the competing combined cycles and gas turbines do. For example, dropping from a 100% load to a 50% load is estimated to increase a combined cycle's heat rate up to 25% and a combustion turbine's heat rate by up to 30%. With their responsiveness maintained, CAES facilities are able to ramp three times as fast as gas combined cycle facilities—which is especially useful on a hot day. However, the real value of CAES is as a mid-merit facility—able to provide this capability at low cost over a much larger period of the year instead of only the 1,000 hours of a peaking gas turbine.

As a sink for power, the decoupled compressors of the CAES facility in the current concepts proposed can play an important role in load management as well, either during off-peak periods (to improve utilization of nearby baseload power facilities by lowering cost and reducing wear and tear) or during normal operation. Although most interest lies in the ability of a CAES facility to absorb large amounts of power during off-peak hours, the ability to reduce or shut off the compression power is another means to rapidly return power to the power grid. It can, therefore, provide a means to respond to very large and rapid shifts in the balance of power output for a region—with additional power output range available from the generator as well. The fixed-speed compression load of a CAES facility is controlled through the use of variable inlet guide vanes and pressure control, which allow the motors to vary their compression load quickly anywhere from 60% to 100% of full load. With a 100-MW compression load, this permits the ability to provide 40 MW of regulation or variation in overall load. (Additional reduction can be achieved through the generating unit—as is the practice with other power plants today.) By controlling the position of these guide vanes, the load absorption on the motor can be controlled, besides controlling the starting and stopping of the unit. In fact, because the compressors and expander turbine train are decoupled, each component can operate at any time.

Although it may seem counterintuitive to operate the compressors while generating power, this gives the facility the flexibility needed to provide a number of ancillary services for the system, such as frequency regulation and load following, voltage control, and contingency reserves. The value of providing these capabilities has always been embedded within the value of existing power generation facilities, but until recently—with the development of an actual market for these capabilities—there has been no price discovery of what the real costs to provide these services are, and how much the market values them. As these markets develop, it is apparent that they are indeed important and valuable, and the faster response capability of a CAES facility gives these units an edge over other providers for these services, increasing the overall value of the facility. Because of this wide range of operations, CAES facilities are also able to benefit wind power. If associated directly with the wind farm, the entire output of wind energy can be structured for the market. Even if a CAES unit is simply located in the same region (because of these facilities' flexible siting arrangements), the excess off-peak power from the wind farms can be put to good use while the local baseload units are allowed to operate more efficiently. As wind energy penetration continues to climb, balancing this variable resource will become more important.

The average round-trip efficiency for existing CAES facilities ranges between 75% and 80%; those in development should achieve upward of 85% efficiency. The charge/discharge ratio for each facility can vary greatly because the charging rate depends on the size of the compressor and the size and pressure of the reservoir versus the output of the generators. Finally, the cycle life of these units has proved to be far greater than similar gas turbines, and the existing facilities are expected to continue operating for some time to come.[5]

Cost issues

System capital costs for next generation CAES facilities are estimated to be approximately $450 per kW according to the *EPRI-DOE Handbook of Energy Storage* (from the Electric Power Research Institute [EPRI]–DOE), with much of the core components benefiting from the overall advancement in turbine technology. Roughly half of this cost is attributable to the power plant components, with the BOP and the cost of the storage reservoir accounting for the remainder of the cost. Construction costs are greatly reduced if an existing salt dome is used rather than having to construct a storage facility. Aquifer facilities are expected to cost less; hard-rock facilities more. Comparatively, modern CAES facilities are expected to have capital costs roughly equal to, or possibly higher than, that of a combined cycle facility on a dollar per kW basis, but—more importantly—will have a much lower cost of energy (produced power) based on favorable off-peak compression power costs.

Installations

Compressed air energy storage technology has been used for nearly 30 years to support electricity generation. Two plants (discussed in the following sections) currently operate daily, with another two under development in the United States alone. Since the Huntorf facility began operation, other proposed facilities have undergone some level of development:

- 500-MW facility in Texas

- 1,050-MW facility in the former Soviet Union

- 300-MW (3 x 100-MW) facility in Israel

- 100-MW facility in Luxembourg

- 25-MW facility in Sesta, Italy

- 35-MW facility in Japan

Additional interest was developed for CAES facilities in South Africa, Korea, and Morocco.

Example—Huntorf, Germany.[6] The world's first commercial CAES facility was the E.ON (German utility) 290-MW Huntorf facility built in Huntorf, Germany, in 1978 to provide originally a blackstart capability for some of the nuclear units near the North Sea. The facility (built by BBC Brown Boveri [now ALSTOM Power]) has successfully run since startup and has averaged 90% availability and 99% reliability. The underground salt cavern reservoir had a volume of 10 million cubic feet, holding air at a pressure up to 1,000 psi. The charging efficiency of the electrically driven compressors for the cavern has averaged 83%, with the cavern normally charged over an eight-hour period to generate 290 MW for up to four hours. With experience, the flexibility of the plant has allowed for additional market uses to be adopted since startup of the facility, increasing its value to both owner and the system operator.

Example—McIntosh, AL.[7] The second CAES facility is the Alabama Electric Company's 110-MW McIntosh facility (power and compressor train from Dresser Rand) in McIntosh, Alabama, developed in 1991 with design assistance from EPRI. The $65-million facility has found its most success operating in the fast load changes during the morning and afternoon peaks. During normal operation, the facility has been started several times a day, ramping quickly to a high load, and after 30 minutes unloaded and returned to hot-standby. A 50-MW compressor is used to fill the 19 million cubic foot underground salt cavern and can provide sufficient air for 26 hours of operation at 100 MW (2,600 MWh). During operation, the air pressure in the chamber draws down from 1,100 psi to 650 psi. The unit is designed to come online within 14 minutes but can start in less than 10 minutes during an emergency (and synchronize within 5 minutes); the facility can run down to 10 MW. It incorporates several improvements over the Huntorf facility, including a recuperator (air-to-air heat exchanger), reducing fuel usage by approximately 25%.

Prospects and challenges

CAES is the only other technology in commercial operation besides PHS able to provide large-scale storage deliverability (more than 100 MW) for use in the wholesale power market. As with pumped-hydro storage, CAES plants have a proven track record, cycling daily and operating efficiently during partial load operations. Most areas in the United States and Europe can support CAES facilities because they possess acceptable geology for the underground storage reservoir. Some areas such as Japan have no such salt layer (strata or domes), prompting them to develop hard-rock mines for the air cavity, like the proposed Norton, Ohio facility. Planned facilities are targeting an availability rate in excess of 95%, with increases in the recuperator efficiency (85% vs. 75%), faster charging times, and a faster airflow from the cavern. These enhancements have the potential to lower the heat rate to near 4,000 Btu per kWh.

Hindering the further deployment of this technology is the perceived unconventional nature of the technology, although CAES relies heavily on commercially available turbine technology. Unlike other power facilities, CAES facilities have significant up-front site development costs associated with prefeasibility tests and underground excavation. For both aquifer and salt-dome reservoir development, preliminary detailed studies of underground conditions are essential and can be very expensive.

However, with the expanded use of underground storage facilities for natural gas storage, these costs and technical requirements are well understood. Development of the currently proposed plants could provide additional operational experience, increasing the confidence in the technology and validation of its economic potential.

Major developers

There are two projects under development for CAES facilities, as described in the following sections.

Example—Iowa Stored Energy Project.[8] The Iowa Stored Energy Project (ISEP) is an integrated power facility, consisting of an 84-MW wind farm connected to a 200-MW CAES unit (Dresser-Rand is providing the power and compressor trains). The $200-million project is set for start-up during the summer of 2006. The compressed air for the CAES plant will be stored in an underground aquifer located 1,200 feet below ground, and it will be able to store the compressed air at 500 psig. Because of the advantageous geology of the area, both air and natural gas will be stored in different strata—providing arbitrage opportunities for gas storage, as well as relieving stress on the local gas pipeline system during peak demand periods. The ISEP will be operated as an intermediate load power plant, producing a mix of wind and power from the CAES facility 12 to 16 hours per day, 5 days per week. During off-peak periods, the ISEP will compress and then store air into an underground aquifer using power from the wind farm (and supplemented by off-peak system power). Overall, the facility is expected to maintain a utilization rate of 50%, with wind power responsible for 33% of the total output. Combining the competitive economics of wind turbines and the CAES facility, the ISEP project is expected to be cost-competitive with other intermediate power plants in the region. Gradually, with the addition of more wind turbine generators and compressors, the facility could evolve into a baseload facility in both scale and capability.

Example—Norton Energy Storage.[9] The CAES Development
Company is developing the Norton Energy Storage 2,700-MW CAES
facility in Norton, Ohio that will use an abandoned limestone mine as
the reservoir. CAES Development Company, a Haddington Ventures,
LLC subsidiary, plans to develop the facility in stages, with the first
300-MW phase expected to cost approximately $200 million. Additional
construction will continue until the facility is developed to its full capacity
of 2,700 MW. ALSTOM Power and MAN TURBO are potential suppliers
for the power and compressor trains. Compressed air will be stored in
an abandoned limestone mine that reaches 2,200 feet deep and will be
capable of storing 338 million cubic feet at 1,500 psi. There is substantial
local support for the CAES project because alternative prospects for the
abandoned mine included developing it to store toxic waste. With a full
charge to the cavern, the full 2,700-MW facility would be able to operate
for an entire 16-hour period, or allow one unit to operate continuously
for 18 days (43,200 MWh).

Related technologies

Other variations of CAES technology are also being developed,
with one in particular termed thermal and compressed air storage
(TACAS). This is essentially a stand-alone and smaller version of CAES
technology. TACAS stores compressed air in conventional high-pressure cylinders, so it can be more easily located on-site at a commercial
or industrial facility. Unlike CAES, which uses natural gas to heat the
compressed air, TACAS systems store thermal energy in a steel thermal
storage unit (TSU) (fig. 3–3).

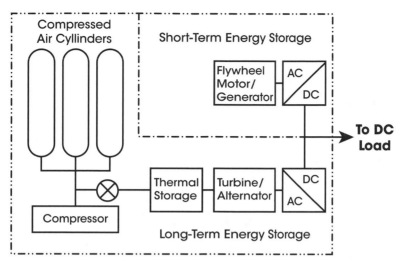

Fig. 3–3. Thermal and compressed air storage (Courtesy of Active Power, Inc.).

TACAS was designed to replace existing battery cabinets in a standard uninterruptible power supply (UPS) system. When low-cost utility power is available, the TACAS system fills its air tanks with its compressor and uses electric resistance heaters to bring the TSU to full operating temperature. After that, TACAS maintains itself in a passive standby mode. A low-cost flywheel is used to maintain power quality and to provide a few seconds of bridging power during operation of the unit. During discharge events, pressurized air is quickly heated in the TSU in route to an expansion turbine. There, the turbine efficiently extracts energy from the high-pressure, high-temperature air to produce power. As this unit is designed to operate indoors, the output temperature is lower than room temperature. When electric utility power returns, the TACAS recharges itself.

TACAS is being developed by Active Power under the trade name of CleanSource XR, in power ranges between 40 and 85 kW. The unit is designed to provide extended run times of 15 minutes or more, with the run-time length determined principally by the volume of air stored on-site. Installed cost and footprint of TACAS are estimated to be comparable to that of flooded batteries for comparable energy storage. Operating expenses (maintenance and replacement costs), however, would be significantly less than conventional UPS batteries, leading to a lower life-cycle cost. As Joe Pinkerton, CEO of Active Power, notes in a personal correspondence, "We believe that the CleanSource XR is the first true minute-for-minute alternative to batteries for the UPS industry. It is ideal for applications where batteries would fail prematurely because of ambient temperatures or heavy cycling loads."

Flow Batteries

Summary

Flow batteries store and release energy through a reversible electrochemical reaction between two electrolytes. There are four types of flow batteries currently in production or in very late stage of development: vanadium redox (fig. 3–4), zinc bromine (fig. 3–5), polysulfide bromide (fig. 3–6), and cerium zinc (fig. 3–7). Flow batteries are typically made from three subsystems (cell stacks, electrolyte tank system, and control system) plus the PCS system. The power and energy ratings of the flow battery are independent of each other. Polysulfide bromide systems are designed at a system level, requiring specific arrays of cell stacks for the particular power rating desired and specific storage tank size for the energy rating desired. Zinc bromine and cerium zinc manufacturers have settled on modular units with fixed power (cell stack dependent) and energy (storage tank dependent) capacities, respectively. Because of this modular design, increasing the capacity of the systems requires discrete units of power and energy capacity. Vanadium redox manufacturers have designed a system using both strategies. During operation, the two electrolytes flow from the separate holding tanks to the cell stack for the reaction, with ions transferred between the two electrolytes across a membrane; after the reaction, the spent electrolytes are returned to the holding tanks. During recharging, this process is reversed. As this technology is highly flexible—both in its physical makeup and its activity potential—it is able to support a wide variety of applications in markets including transmission, retail, and renewable energy.

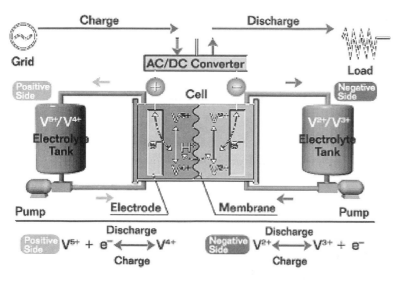

Fig. 3–4. Vanadium redox flow battery (Courtesy of Sumitomo Electric Industries, Ltd. (SEI)).

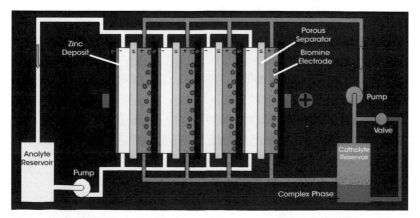

Fig. 3–5. Zinc bromine flow battery (Courtesy of ZBB Energy Corp.).

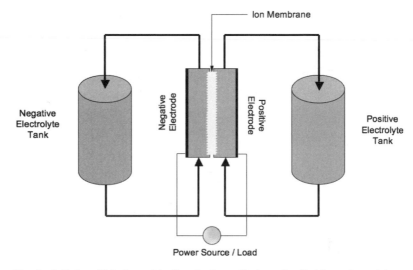

Fig. 3–6. Polysulfide bromide flow battery (Ardour Capital Investments).

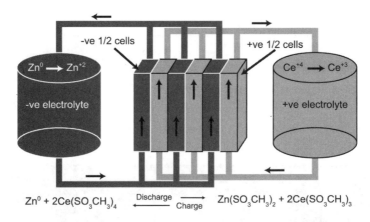

Fig. 3–7. Cerium zinc flow battery (Courtesy of Plurion systems, Inc.).

Historical origins

Vanadium redox. Early work by NASA on iron-chromium (Fe-Cr) flow batteries in the 1970s was the inspiration for research into vanadium redox technology by Professor Maria Skyllas-Kazacos and her team at the

University of New South Wales, in Sydney, Australia, beginning in 1984. Her work focused on vanadium electrolyte stability at high concentration and related processes and technologies. The intellectual property rights surrounding this research and the resulting technology were sold to Pinnacle VRB in 1998. Another line of follow-up work to the NASA research was begun at the Electrotechnical Laboratory (ETL) in Japan in the 1980s. Sumitomo Electric Industries (SEI) then acquired this ETL technology and, in collaboration with the Kansai Electric Power Company, continued to develop the cell stack design and further integrate complete vanadium redox systems (with some additional licensed technology from Pinnacle VRB). Currently SEI is the sole manufacturer of cell stacks, but VRB Power Systems is actively pursuing its own manufacturing capability. VRB Power Systems has recently reached an agreement with SEI to obtain licensing rights to all related intellectual property surrounding vanadium redox technology. Besides the facilities developed by SEI and VRB Power Systems, other development efforts include integration with photovoltaic (PV) arrays (Thai Gypsum Products, later Cellenium Company), an emergency backup system for submarines (Australian Department of Defense), and load leveling (Mitsubishi Chemicals, along with Kashima-Kita Electric Power Corporation).[10]

Zinc bromine. The zinc bromine flow battery technology represents some of the earliest research into any flow battery technology, but any commercial application was difficult due to early poor cycling properties and chemistry challenges. In the 1980s, two veins of research and development started with the hope of using the technology for stationary uses and electric vehicle development. The first area of research began at Gould, Inc., and was later expanded by Energy Research Corporation (which became Evercel). The second area of research centered around Exxon, which developed key components of the technology and licensed it to a number of manufacturers in the United States, Austria, Japan, and Australia. By the mid-1990s, the electric vehicle research focus had waned. Research then focused on stationary power for load-leveling purposes, with a number of stationary pilot projects in the low multi-MW range, and trailer-mounted mobile systems rated at both 200 kW per 400 kWh and 1.8 MW per 1.8 MWh. The first commercialization attempt came through PowerCell Corporation, which in 1998 produced its Powerblock unit that was rated at 100 kW per 100 kWh, but which was shut down by 2002. This event left the zinc bromine flow battery field to ZBB Energy Corporation, a firm that traces its history from Johnson Controls, which licensed the technology from Exxon. Premium Power Corporation, building off the technology developed by PowerCell, has recently begun manufacturing units for sale.[11]

Polysulfide bromide. The original work on polysulfide bromide chemistry was developed by Ralph Zito (United States), an independent inventor who assigned his rights to the technology to Regenesys in the early 1990s. Besides its involvement in and expansion of the body of knowledge on the polysulfide bromide chemistry since the early 1990s, Regenesys Technology had contracted with a number of engineering firms to develop other components of the system. One of these firms, Electrosynthesis, was acquired by Regenesys to assist in the development of the cell stack. After the late 1990s, additional work with Innogy (prior to its acquisition by RWE AG) was undertaken on the first installation-site, Didcot Power Station (100 MWh), before the site selection was changed to Little Barford Power Station facility in the United Kingdom. Besides this power sector installation, the UK's Ministry of Defense had been provided a system to determine the feasibility of including the flow battery technology in an all-electric ship demonstration. The demonstration project was designed to be capable of delivering 1 MW of electricity for periods of 10 minutes under at-sea conditions. VRB Power Systems has recently purchased the technological base of Regenesys Technologies to continue the development of this technology type.[12]

Cerium zinc. Research into cerium zinc flow battery technology began in the early 1990s at Plurion Systems, with the first practical application of this technology developed during 2000 and 2001. Pilot phase demonstration projects are being planned.

Design and operations

Flow batteries comprise three subsystems plus the PCS systems. Since power (MW) and energy (MWh) ratings of flow batteries are independent of each other in the unit's design, increasing the power from the facility necessitates additional cell stacks, whereas expanding the energy capacity of the system is accomplished by expanding the amount of electrolyte stored within the unit. The power (MW) available from each cell is related to the cell voltage (vanadium redox: approximately 1.2 V; polysulfide bromide: approximately 1.5 V; zinc bromine: approximately 1.8 V; and cerium zinc: approximately 2.4 V) and the current density across the membrane. For all flow batteries, each cell is composed of two half cells separated by a microporous membrane (vanadium redox—proton exchange; polysulfide bromide—cation selective); each positive and negative half cell is itself made up of an electrode and a passage for the electrolyte to flow through. Banks of these cells can then be linked together to create a bipolar modular cell stack where the electrodes are shared between the adjacent cells, with the cathode of the first cell becoming the anode of the next cell, and so forth. Linked in series, sufficient cells in these bipolar modules can then form the desired voltage for the cell stack.

The energy storage capacity (MWh) of the facility is determined by the size of the electrolyte holding tank and the concentration of electrolyte used. Normally, rubber-lined or plastic tanks are used to store the electrolyte, with larger systems having flexible geometry to allow for various sizes and shapes. Accompanying the electrolyte tanks is a system of pipes, pumps, and control valves to move the electrolyte from the storage tank to the cell stack and back. Managing all of this is the control system, which manages the activity of the cell and integrates the cell stack electrical connections to the PCS.

Vanadium redox. Energy in vanadium redox flow batteries is stored chemically in two ionic species of vanadium suspended in an aqueous sulfuric acid solution with approximately the same acidity level as that of a lead-acid car battery. During operation (discharge), the two electrolytes flow from the separate holding tanks to the cell stack for the reaction, with ions transferred between the two electrolytes across the proton exchange membrane. The concentration of each ionic form of the vanadium electrolyte changes when the flow battery is discharged, when the potential chemical energy is converted to electrical energy. After the reaction, the spent electrolytes are returned to the holding tanks, but because of the changed chemical nature of the electrolytes, self-separation occurs. During recharging, this process is reversed. Vanadium redox flow batteries operate at normal temperatures, and there is no discharge of the electrolyte solutions from the facility during operation. The electrolytes also have an indefinite life, so they can be used in follow-on installations after removal of the facility.

Zinc bromine. Energy in a zinc bromine flow battery is stored chemically in an aqueous solution of zinc and bromine ions that only differ in their concentration of elemental bromine. During operation (discharge), the electrolytes flow from the separate holding tanks to the cell stack for the reaction. The metallic zinc dissolves into the electrolyte, and zinc ions and bromine ions are allowed to migrate across the microporous polyolefin membrane to the opposite electrolyte, equalizing the charge and converting potential chemical energy to electrical energy. Unlike other flow batteries, the electrodes of the zinc bromine battery cell serve as substrates for the actual chemical reactions. This requires that circulation be maintained to free up surface area for further reaction, and that the battery be fully and regularly discharged to prevent degradation of performance (stripping). During recharging, this process is reversed; zinc is deposited on the negative electrode, and bromine is formed at the positive electrode and remixes with the electrolyte. After the reaction, the spent electrolytes are returned to the holding tanks, but because of the changed chemical nature of the electrolyte, self-separation occurs. Zinc bromine flow batteries can operate in a wide temperature range of 20°C to 50°C. Heat from the operation of the battery can be removed by the use of a small chiller. No electrolyte is discharged from the facility during operation. The electrolytes have an indefinite life, so they can be left in a fully discharged state for extended periods of time with no problem.

Polysulfide bromide. Energy in polysulfide bromide flow batteries is stored chemically in sodium bromide and sodium polysulfide electrolytes. During operation (discharge), the two electrolytes flow from the separate holding tanks to separate halves of the cell for the reaction, separated by the cation-selective membrane—sodium bromide on the positive side and sodium polysulfide on the negative. After the reaction, the spent electrolytes are returned to the holding tanks, but because of the changed chemical nature of the electrolytes, self-separation occurs. During recharging, this process is reversed. Because of the difference in the electrolytes, any failure of a membrane would cause undesirable mixing (precipitating sulfur), so safeguards and monitoring systems are in place to detect a leak. Polysulfide bromide flow batteries operate ideally between 20°C and 40°C, but they are able to tolerate a wider range, with the heat generated during operation removed through the use of a plate cooler. Although there is no discharge of the electrolyte solutions from the facility during operation, trace quantities of bromine and hydrogen gases are produced. The electrolytes present minimal hazards in handling, and from the reaction produce modest amounts of sodium sulfate crystals. Regular maintenance requirements for the facility include a biweekly removal of these sodium sulfate crystal by-products. Other maintenance tasks must be conducted quarterly, including replacement of spent absorbent, and replenishment and maintenance of the electrolytic solution.

Cerium zinc. Energy in cerium zinc flow batteries is stored chemically in zinc and cerium ions suspended in methane sulfonic acid (MSA); the negative half-cell uses a zinc-MSA electrolyte, and the positive half cell uses a cerium-MSA electrolyte. During operation (discharge), the two electrolytes flow from the separate holding tanks to the cell stack for the reaction, with the metallic zinc dissolving into solution, and the cerium ion moving down one in charge as the potential chemical energy is converted to electrical energy. After the reaction, the spent electrolytes are moved to another set of holding tanks to wait recharging of the system. During recharging, this process is reversed. The zinc ions in solution form metallic zinc on the electrode, and the cerium ions move up one in charge. Cerium zinc flow batteries have no discharge of the electrolyte solutions from the facility during operation. The electrolytes are environmentally benign and have an indefinite life. As flow batteries are highly flexible—both in physical makeup and activity potential—these are able to support a wide variety of applications in diverse markets, including transmission, retail, and renewable energy. Units can respond (from standby mode or from charging to discharging) within 20 milliseconds or less, and can ramp up from a full shutdown to full operation within 10 minutes or less. Flow batteries also can sustain elevated output several minutes—some upwards of three times normal—giving the facility a dual-use capability of long-term energy and high response rate with instantaneous discharge capability. In the transmission market, this technology has already proved useful as a multifunctional facility, providing arbitrage, capacity asset deferment, and frequency and voltage regulation. In the retail market, the dual-output capability both acts as an uninterruptible power supply (UPS) to protect against voltage sags, and as a peak-shaving facility in support of energy management strategies. Finally, flow batteries can help integrate renewable energy—both wind and solar—into useful roles in both power grid and remote power applications. Before RWE had stopped development of the technology, Innogy had plans for 12-MW to 100-MW polysulfide bromide flow battery facilities to support wind farms in Denmark in integrating these larger-scale wind energy installations into the power grid.

These wide-ranging applications are supported by the operating characteristics of the different flow battery technologies:

Vanadium redox. Vanadium redox flow batteries have an efficiency approaching 85% (with an estimated 75% AC-to-AC round-trip efficiency, including PCS losses), partially because of the chemical efficiency of the reaction and the small waiting loss while in the electrolyte holding tanks. As the same reversible chemical reaction is responsible for charging and discharging the facility, its charge/discharge ratio capability is 1:1 (although that may not be the norm for each technology). With a cycle life of at least 10,000 charges and discharges, vanadium redox flow batteries have an estimated lifespan of 7 to 15 years, depending on the application and wear of the cell membranes, pumps, and auxiliary components. Although some gradual degradation of the proton exchange membrane occurs (leading to the need for its replacement at the end of its cycle life), no other degradation of capability is expected over the life of the unit.

Zinc bromine. Zinc bromine flow battery cells have an efficiency approaching 80% (with an estimated 70% to 75% AC-to-AC round-trip efficiency, including PCS losses), partially because of the chemical efficiency of the reaction and the small waiting loss while in the electrolyte holding tanks. As the same reversible chemical reaction is responsible for charging and discharging the facility, its charge/discharge ratio is 1:1, although a slower rate is often used to increase efficiency. With a cycle life of at least 2,000 charges and discharges, energy storage facilities based on this technology have an estimated service life of 10 years, depending on the application and wear of the cell membranes, pumps, and auxiliary components. Although some gradual degradation of the microporous membrane occurs (leading to the need for its replacement at the end of its cycle life), no other degradation of capability is expected over the life of the unit, leading to it being capable of fully discharging without damage to the electrolytes or electrodes.

Polysulfide bromide. Polysulfide bromide flow battery cells have an efficiency approaching 75% (with an estimated 65% AC-to-AC round-trip efficiency, including PCS losses), partially because of the chemical efficiency of the reaction and the small waiting loss while in the electrolyte holding tanks. As the same reversible chemical reaction is responsible for charging and discharging the facility, its charge/discharge ratio is 1:1. The life of these facilities is estimated to be 15 years—depending on the application and the wear of the cell membranes, pumps, and auxiliary components. Although some gradual degradation of the cation selective membrane occurs (leading to the need for its replacement at the end of its cycle life), no other degradation of capability is expected over the life of the unit at the facility, leading to it being capable of fully discharging without damage to the electrolytes or electrodes.

Cerium zinc. Cerium zinc flow batteries have a DC–DC efficiency of 70%, partially because of the chemical efficiency of the reaction and the small waiting loss while in the electrolyte holding tanks. As the same reversible chemical reaction is responsible for charging and discharging the facility, its charge/discharge ratio is 1:1. With a high cycle life, energy storage facilities based on this technology have an estimated life of 15 years, depending on the application and the wear of the cell membranes, pumps, and auxiliary components. Although some gradual degradation of the membrane occurs (leading to the need for its replacement at the end of its cycle life), no other degradation of capability is expected over the life of the unit, leading to it being capable of fully discharging without damage to the electrolytes or electrodes.

Cost issues

System costs vary depending on the application desired. According to the *EPRI-DOE Handbook of Energy Storage*, total system costs for a typical multifunctional application of a vanadium redox amount to $1,828 per kW, with 80% of this attributable to the storage module. Also according to the *EPRI-DOE Handbook of Energy Storage*, total system costs for a typical multifunctional application of a polysulfide bromide flow battery are $1,094 per kW, with 80% of this attributable to the storage module. According to this same source, total system costs for a typical multifunctional application of a zinc bromine flow battery amount to $639 per kW, with 60% of this attributable to the storage module. According to Plurion Systems, total system costs for a typical multifunctional application of a cerium zinc flow battery range from $750 per kW to $1,000 per kW, with nearly 50% of this attributable to the storage module. As flow batteries are a recently commercialized product, additional cost reductions are envisioned, which will be primarily driven by modularization and advancements in manufacturing of key components, in addition to scale benefits from larger production runs, rather than any new product breakthrough in material science. It is estimated that the storage modules themselves can be reduced in cost by upward of 25%.[13,14]

Since the power and energy components of flow batteries are decoupled, this affects the cost directly as the units are optimized for one capability over the other. Power-related cost drivers deal with the cost of the cell stack component, such as the electrodes and membranes, and are a large part of the cost of the battery module of the system. Any replacement of these cell stacks is included in the final total ownership of these systems. Incremental energy cost drivers deal with increasing the electrolyte volume and its accompanying piping and storage system. Energy related cost drivers deal with the electrolyte system, including the cost of the electrolyte, the storage tank and piping required, and any control systems required. Often described in terms of dollars per kWh, the electrolyte for the vanadium redox cost is estimated to range from $30 per kWh to $50 per kWh, with total energy storage costs of $300 per kWh to $1,000 per kWh (depending on system design). The costs of the electrolytes for the zinc bromine flow battery are estimated to range from $10 per kWh to $20 per kWh, with total energy storage costs of $400 per kWh (depending on system design). The electrolyte costs for the polysulfide bromide flow battery are estimated to range from

$10 per kWh to $20 per kWh, with total energy storage costs of $160 per kWh to $185 per kWh (depending on system design). Finally, the electrolyte costs for the cerium zinc flow battery are estimated to range from $50 per kWh to $70 per kWh.[15]

The design of the different flow batteries also affects the installation cost. Polysulfide bromide energy storage systems are constructed on-site, leading to somewhat higher EPC (turnkey) costs than purely modular energy storage systems. Since zinc bromine and cerium zinc are modular and not constructed on-site, they have a lower EPC (turnkey) cost than technologies requiring extensive construction on-site. Vanadium redox energy storage systems are designed to be delivered in modular and constructed on-site designs. Another driver for installation costs is the power density of each individual cell. This metric (watts per square meter [W/m^2]) helps to determine the number of cells required for a desired power output, and, hence, the size of the facility.

Installations—vanadium redox

A number of vanadium flow batteries have been operating in a number of countries, including Japan, the United States, South Africa, and Italy.

Example—Totorri Sanyo Electric Company, Ltd.[16] Voltage sags caused by lightning strikes on the local power grid can trip the sensitive equipment in manufacturing facilities for liquid crystal displays—shutting down the processing line and potentially destroying expensive material in the process. SEI was asked by the Tottorri Sanyo Electric Company, Ltd. in Osaka, Japan to develop a UPS solution capable of both protecting the facility and providing flexibility to the facility in managing its energy usage. The company installed a vanadium redox–based flow battery system in April 2001 that provides 3 MW for 1.5 seconds in the UPS role to combat short-term voltage sags. The facility also operates in a peak-shaving role to lower energy costs for the facility by providing 1.5 MW for 1 hour. As opposed to local practices of using distributed generation units as a measure to counter voltage sags, the VRB has no emission—an important benefit as the local government has been pushing for nearby companies to explain and reduce their effects from operations on the local environment. The energy storage facility has operated well, compensating for 22 voltage sags in the first eight months after its installation. This first commercial unit of the VRB is still operating.

Example—University of Stellenbosch, Cape Town, South Africa.[17] A number of utilities worldwide have investigated the use of VRBs. In this example, Eskom, a South African utility, contracted with VRB Power Systems to develop a 250-kW, 520-kWh vanadium bromide battery for installation at the University of Stellenbosch near Cape Town in September 2001. The storage facility installation was a 12-month pilot project and test bed to investigate the technology's UPS capabilities under a number of operational conditions, including power quality, ride-through, and emergency power backup. Besides Eskom and VRB Power Systems, Highveld Vanadium and Steel Corporation was involved in the development of the electrolyte.

Example—PacifiCorp, Moab, UT.[18] In the United States, utilities such as PacifiCorp (a subsidiary of Scottish Power) are looking to use VRBs to postpone the need for upgrading remote power lines. Several potential solutions were examined, including conventional planning options such as line reinforcement, additional power lines, substation upgrades, and increased reactive compensation; however, all were found either impractical (due to environmental restrictions) or very expensive. The use of energy storage as an alternative proved to be the most economically attractive solution, and resulted in the installation of a 250-kW, 2,000-kWh (eight-hour) vanadium redox flow battery energy storage system (VRB-ESS) from VRB Power Systems in late 2003 on a distribution feeder near Moab, Utah. Continuous full power daily cycling operations began in March 2004. To date, feeder voltage deviations have improved by 2%, and feeder power factor improvements have reduced line losses by 40 kW, more than offsetting the parasitic losses of the

VRB-ESS. The unit provides a variety of functions, including peak-shaving, load-following, dynamic frequency control, voltage support, islanded operations, and premium power for industrial and commercial customers on the power line.

Installations—zinc bromine

Although no commercial zinc bromine flow battery facility is currently operating, there have been a number of pilot phase projects both in the United States and Australia.

Example—Melbourne, Victoria, Australia.[19] United Energy Ltd contracted with ZBB Energy to install a 400-kWh zinc bromine flow battery system in a substation to supply a shopping area in suburban Nunawading, Victoria, Australia in November 2001. The facility operates as a test bed for a variety of operations under various seasonal conditions. The system also acts in peak-shaving and system stability roles to defer capacity upgrades on the local distribution system. Financing for the project was provided by United Energy Ltd and the late Energy Research Development Corporation (ERDC). The modular arrangement of the system allows the battery to be isolated from the power grid, which enables ZBB to demonstrate the facility to other utilities.

Example—Detroit Edison, Detroit, MI.[20] One utility with significant experience with energy storage technology, Detroit Edison, installed a 200-kW, 400-kWh zinc bromine flow battery from ZBB Energy for a two-year set of trials in August 2001. Funding for the program was provided through the Energy Storage Systems program of the U.S. DOE's Office of Power Technologies. The zinc bromine flow batteries were chosen over lead-acid batteries for their high energy density (two to three times that of lead-acid batteries), a deeper and far higher cycle life, a faster reaction time, and a rapid recharge capability. Two sets of trials over two years were conducted at Detroit Edison substations in both Akron and Lum, Michigan. Each installation tested different applications.

The Akron installation tested peak shaving for deferral of substation upgrading to alleviate system stress, and the Lum installation provided voltage support for frequent voltage sags. The system was designed to support areas with only seasonal daily peak loads; therefore, it was truck mounted—allowing for quick redeployment of the system to other locations. This capability is important, as areas of voltage instability can quickly grow in areas previously not suffering from these problems because of load growth and usage change by consumers.

Example—Broken Hill, New South Wales.[21] In June 2002, Australian Inland (AI) Energy contracted with ZBB Energy to install a 500-kW zinc bromine flow battery system at a refurbished 20-kW solar generating station for a utility substation in White Cliffs, western New South Wales, Australia. The unit was designed and installed to show the commercial benefits of load management. To support the installation of the unit, the Australian Greenhouse Office (AGO) provided 50% of the total project value of $530,000AU; the balance of funding was provided equally by AI and ZBB Energy.

Installations—polysulfide bromide

Two demonstration facilities were contemplated for Regenesys Technology's polysulfide bromide flow battery; one in the United Kingdom, and one in the United States.

Example—Little Barford Power Station (UK).[22] The first commercial scale polysulfide bromide flow battery demonstration plant from Regenesys was to be located at RWE's Little Barford Power Station in Cambridgeshire, UK. Site preparation work started in August 2000, and the 24,000-cell facility was expected to be in operation by early 2004. The 12-MW, 100-MWh facility was built alongside the Little Barford 680-MW combined cycle power station. It was designed for blackstart capability (40-MWh reserved) and to demonstrate the frequency regulation, voltage control, and arbitrage interaction capability for the surrounding power grid.

Example—Tennessee Valley Authority, Columbus, MS.[23] The second demonstration facility for the polysulfide bromide flow battery technology was at a Tennessee Valley Authority (TVA) substation in Columbus, Mississippi, near the Columbus U.S. Air Force Base. The TVA substation was nearing capacity and would have required $5 million in upgrades to solve the capacity problem using conventional technologies and strategies. The 12-MW, 100-MWh flow battery facility from Regenesys Technologies was originally slated to be online in the first half of 2005. The entire plant would have required two acres of land, containing a building 175 feet long, 65 feet wide, and approximately four stories tall, in addition to the two large electrolyte solution storage tanks. The storage facility's primary role was to provide higher reliability for the Columbus Air Force Base. In addition to supporting the Air Force base, the facility would have served the western edge of the TVA's service territory by providing voltage support, frequency regulation, spinning reserves, and arbitrage opportunities.

Installations—cerium zinc

There are no current commercial or public pilot phase projects of the cerium zinc flow battery technology at this time. Plurion Systems is currently finishing the commercialization of the technology and is readying several pilot projects for installation at demonstration-sites in the near future.

Prospects and challenges

Because of flexibility and their ease of scalability, many utilities are very interested in flow batteries for both supporting transmission and providing enhanced power quality to customers who want premium power for their operations and a means to improve their load profile and, thus, lower their costs. Besides PacifiCorp's installation, both the Municipal Utility Power Company for Boulder City, Nevada, and REMU—a Dutch utility—commissioned studies on flow batteries to study their potential applications for roles such as customer UPS, deferring power grid additions, frequency regulation, and contingency reserve. As these systems have an extremely low marginal cost for additional energy storage, multiple hour capabilities are gaining increasingly competitive cost comparisons versus alternative storage technologies. As Gary Colello, CEO of Premium Power Corporation notes in a personal correspondence, "Environmentally friendly Zinc-Flow energy storage is adaptable to a variety of target markets and applications including telecommunications, electric utilities, datacenters, and renewables such as wind and solar. In addition to meeting a full spectrum of scalable AC and DC output power requirements, Zinc-Flow energy storage is fully repeatable for thousands of cycles and provides 10 times the energy storage when compared to lead-acid batteries. Premium Power solutions are designed for turnkey outdoor or indoor placement and installation can be accomplished in under 4 hours. With only minimal required maintenance, Zinc-Flow provides reliable operation for up to 30 years." VRB Power Systems has recently purchased the technological base of Regenesys Technologies to continue the development of the polysulfide bromide flow battery technology. The polysulfide bromide flow battery in particular is of interest to those wanting larger-scale units because of its potential to scale easily into the multi-MW range.

Although continued interest exists, failures of companies such as PowerCell (zinc bromine) and Regenesys Technologies (polysulfide bromide) continue to cast doubt over all flow battery technology. Additional facility market penetration will help both in the manufacturing of lower-cost cell stacks and on-site facility construction knowledge. Other possible advancements such as improvement of the concentration level of the electrolyte would reduce storage requirements (volume for a given energy level), reducing costs for the auxiliary systems. Each flow battery technology also many times has its own hurdle. Vanadium redox flow batteries have the lowest power density and require the largest number of cells to attain the same rated system power output. Plurion Systems (the only manufacturer of cerium zinc flow batteries) has yet to produce a commercial product. Greater commercial success is also required of all of these technologies to establish a broader track record for these technologies. As multifunctional facilities, flow batteries must many times face single-purpose competitors that perform a particular market role more cheaply. Storage technologies that provide multiple benefits to multiple beneficiaries rarely are the least expensive for any one role. What is not more inexpensive, however, is the current (but often overlooked) requirement to buy all the requisite competing technologies to match the flexibility of these systems. That message must be conveyed to potential customers more forcefully for these types of systems to be successful.

Major developers

Vanadium redox. Two major developers are currently producing commercial scale vanadium-based flow battery systems: Sumitomo Electric Industries, Ltd (SEI) and VRB Power Systems. Cellenium Company, Ltd. has also undertaken some development work toward providing vanadium flow cell batteries but has not entered into commercial-scale production as of this writing.

Zinc bromine. There are two active developers of the zinc bromine flow batteries: ZBB Energy Corporation and Premium Power Corporation. Premium Power purchased the technology of PowerCell and has recently begun to produce units for sale.

Polysulfide bromide. VRB Power Systems has recently purchased the technological base of Regenesys to continue the development of the polysulfide bromide flow battery technology. Previously, Regenesys Technologies, Ltd., a subsidiary of RWE AG (formerly Innogy, formerly National Power) was the only developer of this technology.

Cerium zinc. The only developer of the cerium zinc flow battery technology is Plurion Systems, Inc.

Sodium Sulfur Battery

Summary

The sodium sulfur battery's (NAS) battery cell is a cylindrical electrochemical cell with a molten-sodium negative electrode in the center and a molten-sulfur positive electrode on the outside, separated from the negative electrode by the β-alumina solid electrolyte (fig. 3–8). As a cell is discharged, sodium at the negative electrode discharges electrons, and sodium ions pass through the β-alumina electrolyte to the positive electrode, where they react with sulfur to form sodium polysulfide. When a cell is charged, this reaction is reversed; the sodium polysulfide at the positive electrode decomposes, and sodium ions return to the positive electrode. Average round-trip efficiency is 89% for the storage module. Prospects for this technology are promising in the retail market for energy management and power quality. One significant benefit of this technology is its ability to provide output in either a long-term or pulse mode. To date, the technology has been well received in Japan, and international evaluation is underway, with one unit operating in the United States, and additional market development work in progress in Europe and Southeast Asia.

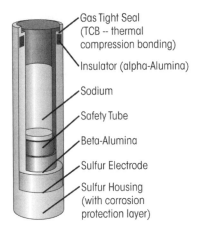

- **Low resistance, high efficiency due to**
  - Beta Alumina tube
  - Sulfur electrode design

- **High durability due to**
  - Corrosion protection layer
  - Sulfur electrode design

- **High energy density due to**
  - Cell properties and design

- **Intrinsic safety due to**
  - Incorporation of safety tube

Gas Tight Seal (TCB -- thermal compression bonding)

Insulator (alpha-Alumina)

Sodium

Safety Tube

Beta-Alumina

Sulfur Electrode

Sulfur Housing (with corrosion protection layer)

Fig. 3–8. Sodium sulfur (NAS) battery (Courtesy of NGK Insulators, Ltd.).

Historical origins

The Ford Motor Company conducted early development of the NAS battery in the 1960s for electric vehicles, closely followed by ABB in Europe and the New Energy and Industrial Technology Development Organization in Japan. General Electric (GE) undertook other early research, this time into stationary applications, during the late 1970s. Starting in the early 1980s, the Tokyo Electric Power Company (TEPCO) began a joint program with NGK Insulators to develop the technology for load leveling at substations and customer sites (initial target, 2-MW,

16-MWh units). With demand for electricity increasing rapidly and a poor system load factor, TEPCO needed to develop a means to entice its customers to help improve the operation of the system—and save themselves money at the same time. This need led NGK to establish a research program with ABB in the late 1980s, which resulted in a licensing agreement for ABB NAS technology in 1998. NGK's NAS battery development program has progressed through three significant stages:

1. β-alumina tube improvements

2. NAS cell and module design

3. Optimization of energy density, performance, and life parameters

By the early 1990s, NGK and TEPCO began installing the first demonstration programs at customer sites. By 2002, the technology became commercial in Japan, and the first demonstration unit was introduced into the United States at an American Electric Power (AEP) facility.[24]

Design and operations

The NAS battery cell is a cylindrical electrochemical cell with a molten-sodium negative electrode in the center and a molten-sulfur positive electrode on the outside, separated from the negative electrode by the β-alumina solid electrolyte. Only the positive sodium ions are allowed to migrate through the electrolyte to react with the sulfur to form sodium polysulfides. The battery must be maintained at an operating temperature of 320°C to 340°C to promote this reaction. As a cell is discharged, sodium at the negative electrode discharges electrons, and sodium ions pass through the β-alumina electrolyte to the positive electrode, where they react with sulfur to form sodium polysulfide. When a cell is charged, this reaction is reversed; the sodium polysulfide at the positive electrode decomposes, and sodium ions return to the negative electrode. Each NAS cell operates at approximately 2 V, and cells are configured in series/parallel arrays to provide the specified voltage and energy storage capability (energy density three times that of a lead-acid battery). Each cell has no self-discharge or memory effect common among chemical batteries. All of the individual cells are then housed in a thermally insulated enclosure to form a typical NAS battery module of 50 kW, 360 kWh or 50 kW, 430 kWh. Through the cell arrangement, these modules can then be optimized for power quality (power) or peak shaving (energy). Because of the internal heaters, the battery module is insensitive to ambient temperatures and can therefore be easily placed outdoors.

NAS battery systems can provide power either in a single, continuous discharge or in a larger, but shorter, pulse of power. This flexibility allows the unit to perform a variety of market applications. Utilities including TEPCO and AEP have evaluated the system for use at utility substations to defer transmission upgrades. NAS batteries have also been evaluated and deployed for use in energy management (peak shaving) and power quality in the retail energy market. In the peak-shaving role, the unit is called on to provide long-duration discharges for groups with large loads, such as office buildings, factories, shopping centers, and schools. Because of the multi-MW potential size of these units, the incumbent utility also benefits from an improved load factor among its larger customers. The short-term pulse capability of the system also enables the unit to counteract power quality disturbances. This pulse-power capability is rated at five times the unit's continuous rating. The ability to inject power for short periods is useful to mitigate power disturbances such as voltage sags or momentary outages. For example, the ability to provide power for 30 seconds is useful because it is a sufficient amount to address more than 98% of all power quality events and to support a transition to a backup generator. This pulse-power capability is also available even if the unit is currently in the middle of a long-term discharge for peak shaving. With a growing body of operational knowledge with the NAS batteries, additional market roles are continuing to be evaluated, including using the unit to stabilize the variable energy output of wind turbines to improve the integration of wind power into the power grid.

These wide-ranging applications are supported by NAS battery systems because of their strong operating characteristics. NAS batteries have a high round-trip efficiency and maintain an 89% efficiency rate during normal operation. With minimal cell degradation, the cycle life is also far higher than that of other chemical batteries—at 100% depth of discharge (DOD), the NAS battery has a cycle life of approximately 2,500 cycles. As with other chemical batteries, shallower discharges provide for a longer cycle life. At 90% DOD, the unit has a 4,500 cycle life; at 65% DOD, the unit has a 6,500 cycle life; and at a 20% DOD, the unit has a 40,000 cycle life. Through an expected 15-year operational life, the degree of degradation for the NAS battery cell is highly related to the corrosion of the sulfur electrode. Whereas both power quality and peak-shaving modules charge at 50 kW, each variant discharges at a different rate, thereby giving a different charge/discharge ratio depending on design and market role. The peak-shaving NAS battery module can discharge at up to 100 kW for 15 minutes, in addition to discharging at lower power levels for longer time periods. Conversely, the power quality NAS battery module can discharge at up to 250 kW for 30 seconds or longer, in addition to discharging at lower power levels for longer periods of time.[25]

Cost issues

According to the DOE-EPRI *Handbook of Energy Storage*, total system costs for a typical multifunctional NAS battery are $810 per kW, with 60% of this attributable to the battery module. As this is a recently commercialized product, additional cost reductions are envisioned, which will primarily be driven by manufacturing advancements and scale benefits from larger production runs, rather than any new product breakthrough in material science. It is estimated that the battery modules themselves can be reduced in cost by upward of 33%.[26]

Installations

By early 2004, more than 50 MW of NAS battery systems were operating at a number of locations, primarily in Japan, but also in countries like the United States, where evaluation studies are underway. In total, more than 88 of these systems were installed between 1992 and 2004, totaling more than 66 MW of capacity. Existing units range from 2,000-kW, 14,400-kWh units for large industrial factories and semiconductor manufacturing factories to 100-kW, 720-kWh units for hospitals and amusement facilities for emergency power or peak-shaving strategies. The largest installation to date is an 8-MW, 64-MWh facility installed by TEPCO at a water treatment facility in Morigasaki, Japan, in April 2004.

Example—Akiruno, Japan.[27] A Fujitsu semiconductor factory in Akiruno, Japan, suffered periodic voltage sags sufficient to cause major processing disruptions (and damage to plant equipment) from lighting strikes. In fact, during the year prior to finding a solution to the problem, five major lightning strikes occurred. Through working with the local utility, TEPCO, the customer chose to install an NAS battery system from NGK Insulators that began operation July 15, 2002. The NAS battery system is connected with the power grid of the customer's substation at 6.6 kV and has operated nearly every day in peak-shaving mode, while providing enhanced power quality to the mission-critical components of the semiconductor factory. For peak shaving, the unit is capable of providing 1 MW for 7.2 hours, and has been able to be retasked to alternative operation schedules according to the factory's changing demand pattern. For power quality protection, the NAS battery provides up to 3 MVA for 13.5 seconds, with the transition time between peak shaving to the pulse power output capable within

20 milliseconds. Since installation, four voltage sags caused by lightning strikes have caused the NAS battery system to provide power quality for the semiconductor facility. Although two of these power quality events were multiple lightning strike events that occurred many seconds apart, the NAS battery system was able to provide sufficient protection during all events that no mission-critical component was damaged or tripped because of the voltage sag.

Example—boat racing facility, Japan.[28] In another example, an NAS battery system was chosen to provide higher reliability and lower cost power for lighting at a public gambling facility in Japan. At the 9,000-person facility, boat racing is a very popular pastime, and night races, held 100 times per year, are especially popular. The facility's owner contacted Kyushu Electric Power Company to find ways to both improve the reliability of the lighting system and lower the cost of service. The large lighting system needed to support the races was a significant portion of the facility's 4,400-kW peak load and total energy usage of 5,200 MWh. A 2-MW, 14.4-MWh NAS battery system was chosen, which was owned and maintained by Kyushu Electric and leased to the owners of the boat racing facility. Other options such as distributed generation (DG) or other battery systems were not chosen because the NAS battery facility required less operation and maintenance, was quieter as it has no moving parts, and, with no fuel used, had much less environmental impact. After installation, the peak load of the facility was reduced more than 40% to 2,600 kW. This reduction was a significant factor in the $1,170,000 annual cost of electric service being reduced 28% to $842,000.

Example—Gahanna, Ohio.[29] AEP installed the first NAS battery system in North America at an office complex in Gahanna, Ohio, in September 2002. The system is rated for 100 kW, 720 kWh in the peak-shaving mode (7 hours), with a 30-second 500-kW pulse capability for power quality applications (ABB provided the system's power electronics). The DOE and EPRI provided assessment during the two-year pilot project phase of the facility's peak shaving and power quality operational performance to evaluate its economic costs and benefits for a variety of customer and utility applications. During the evaluation phase, the unit operated in a peak-shaving mode five days per week, shifting a total of 15 MWh per month. The unit was also subjected to 200 simulated power quality events, besides experiencing 90 actual power quality events.

Prospects and challenges

Prospects for this technology show promise both in existing markets and new ones. The NAS battery compares especially favorably with existing lead-acid batteries. As the system encompasses additional capabilities that lead-acid batteries cannot accommodate, these facilities can take on additional market roles in harsher environments. Specifically, NAS batteries have:

- Three times the energy density of lead-acid batteries (more efficient use of space)

- Longer life span because of high cycle ability

- Ease of maintenance (periodic inspection and cleaning) for the sealed battery modules, as there are no moving parts such as pumps or heat exchangers to the system

Because of these characteristics and the MW scalability of these units, they possess strong environmental benefits compared to on-site generation of peak shaving, as the NAS battery has no emissions during operation. This aspect is especially important in congested urban environments where there are strict emission rules.

To date, the technology has been well received in Japan, and international evaluation is underway with one unit operating in the United States and additional market development work underway in Europe and Southeast Asia. As Ted Takayama, Manager of Business

Planning for NGK's NAS Battery Division, notes, "Utility-scale energy storage with the NAS battery is a commercial reality in Japan and ready for introduction to the global market. Through extensive testing and precommercial demonstration projects with the Tokyo Electric Power Company, high reliability and safety have been confirmed. NGK's commercial scale manufacturing facilities are now in operation with plans for expansion to meet the growing market."

Barriers to low-cost and large-scale manufacturing have been largely eliminated as the material of the batteries is inexpensive and abundant. In addition, with a nod toward the life cycle responsibility of modern manufacturing in Europe and Japan, dismantling the batteries has been made easier, as more than 99% of the battery material can be recycled.

Challenges for this battery system include those similar to other energy management strategies, which rely on arbitrage—the users' energy cost must be high enough to warrant the installation of the unit. However, the pulse power capability of the system for power quality solutions provides an additional market use of the unit, helping to make it more cost-effective to customers in lower energy cost areas that need more than simply lower energy costs to warrant the installation of the unit.

Going forward, this technology will continue to focus on retail energy management and power quality applications, but as transmission and distribution upgrades become increasingly difficult everywhere, deferring utility transmission capacity upgrades will become a larger target for development. To that end, in 2004 the Japanese government began providing research and development funding for advanced load-leveling equipment, such as an efficient heat storage facility and the NAS battery system, to pursue an emerging national load-leveling policy.

Major developers

TEPCO and NGK Insulators of Japan are the only developers of NAS batteries today (commercial production began in the spring of 2002), with a current manufacturing capacity of 65 MW, with plans for a 200-MW capacity in a few years.

Lead-Acid Battery

Summary

Research and development of lead-acid (LA) battery technology has been ongoing for more than 140 years. The two predominant types of LA batteries are flooded (or vented) and valve regulated (or sealed) (fig. 3–9). Flooded lead-acid batteries are used in three applications: starting and ignition, deep cycle, and industrial uses, while VRLA are used in applications such as industrial tools and backup power. The electrodes in LA batteries are used both for part of the chemical reaction and for storing the results of the chemical reactions on their surfaces. Therefore, both the energy storage capacity and power rating are based on the size and geometry of the electrodes. Because of their low cost and reliability, LA batteries remain a favorite for a wide variety of market applications in the transmission, retail, and renewable energy markets. Although the possibility exists to use them in a variety of applications, environmental and operational effects curtail the list of applications truly available to them. However, LA batteries will remain an important energy storage technology in a number of existing market applications for the foreseeable future; they will always be the low-cost option for less-taxing applications in the UPS, telecommunication, and remote/off-grid renewable markets. However, prospects for this technology in the expanding role of energy storage technologies are limited.

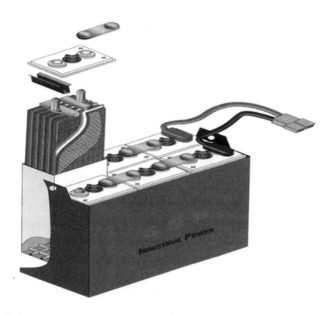

Fig. 3–9. Lead-acid (LA) battery (Courtesy of GNB Industrial Power, a division of Exide Technologies).

Historical origins

The first practical LA battery was developed by Gaston Plante, a French scientist, in 1859. By the 1870s, this technology was integrated into early power-delivery systems to provide load-leveling capability to meet peak demands. Although rapidly expanding power stations and transmission systems made the need for prepositioned energy sources less necessary, the early 20th century saw increased use of LA batteries to power early electric automobiles. As the internal combustion engine quickly took over the market for engines, LA batteries found a niche in the market as the starter for electric motors by the 1920s. Also by the 1920s, LA batteries had become widespread in the utility market for standby power systems in power plants and substations. Research into less expensive and more efficient manufacturing of a more reliable flooded LA battery created new applications, such as the energy storage component of a UPS system. By the 1970s, the valve-regulated lead-acid (VRLA) battery was developed, with a lower manufacturing cost and easier maintenance, but a shorter life. Since then, both flooded and VRLA batteries have found uses in the stationary market. By 2001, the total world LA battery market amounted to $15 billion (wholesale), with the stationary market accounting for approximately 8% of this value.[30]

Design and operations

Two types dominate LA battery design: flooded (or vented), and valve-regulated lead-acid battery—VRLA (or sealed). Flooded LA batteries are used in three areas:

1. Starting and ignition, where they provide a short burst of strong power to start an engine

2. Deep cycle, where they provide a low, but steady level of power for a longer period of time than a starting battery

3. Industrial, where they provide low, steady power over a much longer period of time than a typical deep cycle battery

A flooded LA battery consists of a positive electrode comprised of lead dioxide and a negative electrode comprised of lead; to improve the performance of the battery, pure lead is not used, but rather an alloy, such as lead-antimony. The electrodes are then immersed in liquid electrolyte, a dilute solution of sulfuric acid and water (35% acid/65% water).

VRLAs are used in a number of applications, including industrial tools and backup power. Unlike flooded LA batteries, these batteries are sealed to prevent water loss or release of gas during charging. Although designed to be a longer-life replacement for flooded LA batteries in applications such as UPS installations, VRLA batteries are actually far more susceptible to variations in temperature and over/undercharging that affect the life of the battery—giving them a lower operating life than flooded LA batteries. Each VRLA cell is constructed from positive and negative electrodes that are separated by a porous material, although the electrolyte is immobilized into an absorbent separator or gel. VRLA batteries also contain a resealable vent to allow off-gassing if the battery is overcharged; this limits the rate at which VRLA batteries can be charged. For both designs, the cells in a battery are connected in parallel series to obtain the required current and voltage for the application. As each battery has slightly different operating characteristics, monitoring systems are needed to both balance the current flow and signal if a battery is failing or fails (because of an internal fault) to prevent neighboring batteries from damage.

The electrodes in LA batteries are used for part of the chemical reaction and for storing the results of the chemical reactions on their surfaces. Therefore, both the energy storage capacity and the power rating are based on the size and geometry of the electrodes. A higher power rating requires a larger surface area for each electrode, often leading to more and thinner plates in a battery. However, the energy storage capability is based on the mass of the plate, leading to fewer and thicker plates (in a given sized battery). Typically, LA batteries have a cell voltage of 2 V, but as the battery discharges, the voltage drops slowly. During discharge, the positive electrode (lead dioxide) frees lead ions into the electrolyte solution, while lead sulfate forms on the negative electrode (lead), as the sulfuric acid turns into water. Reversing the electrical charge through the system recharges the battery. When the cell is being recharged, the chemical reactions are reversed, restoring the battery to its original condition. Because of off-gassing, flooded LA batteries require far more monitoring and attention (adding water) than VRLA batteries. In addition, the emission of acid fumes from flooded LA batteries can cause corrosion of surrounding metal supports unless vented to the outside.[31]

Because of their reliability and low cost, LA batteries remain a favorite for a wide variety of market applications in the transmission, retail, and renewable energy markets. These batteries can respond within milliseconds and provide full power instantaneously. In the transmission market, LA batteries have already proved useful as a multifunctional facility, providing capacity asset deferment along with frequency and voltage regulation. However, these demonstration projects showed that larger MW-scale facilities can have operational difficulty, although smaller-scale units have proved useful in niche locations. In the retail market, the ability to produce double the normal output levels for short periods allows these units to both provide commercial and industrial facilities power quality protection against voltage sags and be a ride-through resource. Some groups are also investigating combining LA batteries with shorter-duration storage technologies to provide a longer emergency power capability. Another important use of LA batteries is to provide standby power for substations, power plants, and the telecommunication industry. However, peak shaving for commercial and industrial facilities would shorten the battery life because of the higher cyclic requirements of that application, so integrating them into an energy management strategy is generally avoided. Finally, LA batteries are used extensively in remote renewable energy installations—especially solar—to increase the value of the power by providing dispatchability.

Although the possibility exists to use LA batteries in a variety of applications, environmental and operational effects curtail the applications truly available to them. The average DC–DC round-trip efficiency of an LA battery is 75% to 85% during normal operation. Generally, an LA battery has a useful life of approximately five years under normal operating conditions, which corresponds to a cycle limit of 250–1,000 charge/discharge cycles, with VRLA batteries having a much lower cycle life than vented LA batteries. The cycle life of an LA battery can be significantly degraded because of temperature, depth of discharge (DOD), fast charge/discharge cycling, and other factors. Temperature may be one of the most important aspects affecting the cycle life of an LA battery. The primary operating temperature of an LA battery is roughly 80°F, but operating the battery 40 or more degrees above this point can cut the life of the battery by 50%. (Many UPS systems have minimal space-conditioning, leading to early failure.) The DOD also affects the cycle life of an LA battery—with the deeper average discharge corresponding to a shorter life. Normally, LA batteries for UPS and other energy storage systems are designed for steady, deep cycle discharges of 50% to 80%; above this, the effect becomes more pronounced. Finally, LA batteries take far longer to charge than discharge, having an effective charge-to-discharge ratio of 5:1 or more to prevent damage to the cell. Faster charging, although possible, will lower the life of the battery; if charged at more than 2.4 V, gassing (electrolysis) occurs (this also occurs during overcharging), producing oxygen and hydrogen and ultimately damaging the cells.[32]

Cost issues

Generally, LA batteries are the cheapest energy storage technology choice for most retail and some transmission applications. According to the *EPRI-DOE Handbook of Energy Storage*, a multi-MW, multifunctional LA battery system has a system cost of $580 per kW, with a little more than 50% from the battery module. As these are mature technologies, there is no real expectation for significant cost reductions over the near term from technology or manufacturing advancements.[33]

However, cost comparisons should incorporate the real-world operating requirements and conditions that often shorten the expected life of LA battery units—driving up their actual total ownership cost (in particular, high cycle, deep discharge conditions), even to the point where they become more expensive than alternative energy storage technologies. Design issues of matching capabilities to applications are also important because the power and energy components of this technology are linked as each cell has a specific aspect of each. Therefore, the scale (and hence cost) of an LA battery system relies to a great extent on the required number of battery strings. Each string requires an accompanying monitoring system to prevent failure of one battery from damaging other batteries within the string. Smaller groups of LA batteries for UPS systems can be modular, but larger, multifunctional facilities require some construction on-site for the housing, monitoring, and space conditioning system. Although the space conditioning system is looked on for cost savings, failure to provide an environment that does not stress the batteries will greatly reduce their useful life.

Installations

Besides numerous UPS installations, LA batteries were the heart of many of the early large-scale multifunctional energy storage facilities, such as the 1986 BEWAG 8.5-MW, 8.5-MWh battery facility located in Berlin, Germany.

Example—Vernon, CA.[34] An example of a large LA battery UPS system is the 3-MW, 4.5-MWh VRLA UPS system that was installed at an LA battery recycling plant in Vernon, California. This facility was designed to provide peak demand reduction and uninterruptible power. For a number of years after its installation in 1996, the system greatly reduced the power quality events that had previously resulted in increased lead emissions and costly noncompliance fines.

Example—Chino, CA.[35] Southern California Edison (now a subsidiary of Edison International) installed a 10-MW, 40-MWh vented LA battery system during 1988 at a substation in Chino, California. The system was installed as a pilot facility to evaluate battery systems. Its primary role was to evaluate load-leveling operations of a battery energy storage facility. Secondarily, the facility evaluated load following, transmission asset deferral, economic dispatch, frequency regulation, voltage control, and blackstart operation capability. The facility operated successfully through 1997.

Example—Sabana Llana, Puerto Rico.[36] An early example was the Puerto Rico Electric Power Authority's 20-MW, 14-MWh (40 minute) LA battery storage plant at its Sabana Llana substation, which was installed in 1994. This $21-million facility successfully provided spinning reserves, regulation, and voltage control services for the island during normal operation. Because of the small size of the island's power grid, these capabilities were always in short supply. These capabilities proved extremely useful during the fall of 1998 in the aftermath of Hurricane George, when unreliable power supplies and transmission reliability caused a number of load-shedding events and increased spinning reserve requirements—requirements that the battery facility reduced during this time of scant resources. Although the plant operated successfully for a number of years, deteriorating batteries caused the premature closure of the facility in 1999.

Example—Metlakatla, AK.[37] In 1997, Metlakatla Power & Light in Metlakatla, Alaska—a small village on an island in the southeast portion of Alaska—installed a 1.2-MW, 1.4-MWh valve-regulated LA battery system from GNB Industrial Power/Exide. The facility was designed to have a 20-year life span. This Metlakatla facility is a well-known example of how a battery energy storage facility can help integrate renewable energy resources into a power grid, and how a storage facility can become an integral part of an island grid's balancing system. Here, the full cost of the $1.5-million energy storage facility was recovered within three years because the storage unit improved overall power quality and reduced the operating costs of the existing generating assets. There were both direct financial and nonfinancial benefits to using the system. The system matched generation needs very well, almost totally replacing a peaking diesel unit that was costing $400,000 per year in fuel costs alone. In addition, the energy storage facility greatly reduced the environmental risks to the community, as diesel emissions and noise were greatly reduced (or eliminated).

Prospects and challenges

LA batteries will remain an important energy storage technology in a number of existing market applications for the foreseeable future; they will always be the low-cost option for less taxing applications in the UPS, telecommunication, and remote and off-grid renewable markets. However, prospects for this technology's usefulness in the expanding role of energy storage technologies are limited because the requirements of many of these applications would significantly curtail the life of the unit. Although mature, further development will continue (primarily through using different alloys with lead for the electrodes) to reduce maintenance and extend life through reducing corrosion, plate buckling, and off-gassing.

Unfortunately, a number of challenges inherent in the technology will continue to restrain further deployment of LA batteries for stationary applications in the electric utility market. Recent advances in the basic technology continue to fall short of the breakthrough desired. This lack of technological innovations restrains significant additional research and growth for this technology, and encourages research into competing energy storage technologies. For these reasons, the roles currently assigned to LA batteries will be the only practical ones available; large-scale, multi-MW systems are impractical because of short battery life and the extensive monitoring and environmental controls required. Although the markets dominated by LA batteries will continue to grow, a number of other energy storage technologies will continue to specifically target these established LA battery markets for growth. As environmental concerns about lead continue to grow, standards and regulations concerning the manufacturing, handling, and disposal of these batteries will continue to increase.[38]

Major developers

There are a number of manufacturers of LA batteries for a variety of applications. Some of the larger manufacturers include C&D Technologies, GNB Industrial Power/Exide, East Penn Manufacturing Company, Gill Batteries (a division of Teledyne Continental Motors), Optima Batteries (a Johnson Controls, Inc, company), Trojan Battery Company, and Crown Battery Manufacturing Company.

Nickel-Cadmium Battery

Summary

Swedish scientist Waldmar Jungner developed the first nickel-cadmium (NiCd) battery (the pocket-plate style) in 1899. Three designs dominate nickel-cadmium battery design: pocket plated, sintered plated, and sealed. A typical design can be seen in figure 3–10. Pocket-plated NiCd batteries are used extensively in markets such as industrial and standby power, where ruggedness and durability are important. Sintered NiCd batteries dominate in markets such as for starting aircraft and diesel engines, where high energy per weight and volume are important. Finally, sealed NiCd batteries are used commonly in commercial electronic products, where lightweight, portable, and rechargeable power is important. The reliability of NiCd batteries makes them a favorite for a wide variety of market applications in the transmission, retail, and renewable energy markets. For many reserve power applications, NiCd batteries are one of the least expensive energy storage technology choices for most applications in the retail and some transmission applications. Nickel-cadmium batteries will remain an important energy storage technology in a number of existing market applications for the foreseeable future. Although slightly more expensive than LA batteries, their operating capabilities and excellent reliability span a wider operations envelope and longer cycle life, allowing for lower ownership costs. Unfortunately, a number of challenges continue to restrain significant further deployment of NiCd batteries for stationary applications in the electric utility market.

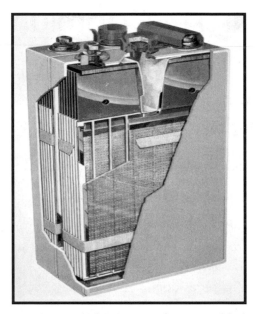

Fig. 3–10. Nickel cadmium (NiCd) battery (Courtesy of Saft).

Historical origins

Waldmar Jungner's pocket-plate NiCd battery had few applications because of the high cost and difficulty in manufacture. For this reason, much development activity for nickel-based batteries did not occur until after Thomas Edison designed the nickel-iron battery. In 1932, Shlecht and Ackermann invented the sintered plate design, which was thinner with a higher surface area, allowing much higher load currents and improved longevity. By 1946, Neumann developed the sealed nickel cadmium cells, with early applications in the military market. Development toward improving manufacturing and design continued, focusing on improving capabilities and lowering the cost of the unit. Pocket-plate designs still dominate for ruggedness and durability in market applications, whereas sintered-plated designs dominate where higher energy per weight and volume are important. Nickel-cadmium batteries are found throughout the rail industry (40% of all NiCd batteries are destined for this market), military uses, space applications, and standby power in the industrial and electric power industry. By the 1970s, development of the technology had continued so that NiCd batteries became widely used in the portable power market for consumer electronics.[39]

Design and operations

The three dominant NiCd battery designs are pocket plated, sintered plated, and sealed. Pocket-plated NiCd batteries are used extensively in markets such as industrial and standby power, where ruggedness and durability are important. Sintered-plated NiCd batteries dominate in markets such as aircraft and diesel engine starters, where higher energy per weight and volume are important. Sealed NiCd batteries are used commonly in commercial electronics products, where lightweight, portable, and rechargeable power is important.

A NiCd battery cell consists of a positive electrode comprised of nickel hydroxide and a negative electrode comprised of metallic cadmium commonly divided by a nylon separator. The electrolyte is aqueous potassium hydroxide and is not significantly changed during the reaction, with the cell voltage of the typical NiCd battery being 1.2 V. During discharge, nickel oxyhydroxide combines with water to produce nickel hydroxide and a hydroxide ion, and cadmium hydroxide is produced at the negative electrode. Reversing the electrical charge through the system recharges the battery. When the cell is being recharged, the chemical reactions are reversed, restoring the battery to its original condition. During charging, some oxygen is produced at the positive electrode and some hydrogen at the negative electrode, requiring some venting and occasional water addition, but far less than that required of an LA battery.[40]

Because of the reliability of NiCd batteries, they remain a favorite for a wide variety of market applications in the transmission, retail, and renewable energy markets. These batteries can respond within milliseconds and provide full power instantaneously. In the transmission market,

NiCd batteries have recently been installed in the largest multifunctional battery energy storage facility—the 40-MW Golden Valley Electricity Association facility—to provide much-needed frequency and voltage regulation on that utility's system. As with LA batteries, however, only experience will prove if this technology is capable of wider deployment in MW-scale facilities for this market.

In the retail market, NiCd batteries have proved popular with commercial and industrial facilities that need power quality protection against voltage sags and a ride-through resource to on-site generation for firms looking at total ownership costs instead of simple up-front costs. These batteries are also finding a growing market as a replacement for LA batteries to provide standby power in harsh environments such as power plants. Although possessing a better cycle life than LA batteries, peak shaving for commercial and industrial facilities would shorten the NiCd battery life because of the higher cyclic requirements of that application, so integrating them into an energy management strategy is generally avoided. Nickel cadmium batteries are also finding many new applications for remote renewable energy installations—especially solar—where reliability and tolerance to heat is required (something common where solar energy intensity is strong).

These wide-ranging applications are supported by nickel cadmium batteries' useful operating capabilities. Unfortunately, the average DC–DC round-trip efficiency of a NiCd battery is 60% to 70% during normal operation—somewhat lower than LA batteries. However, the expected life of these batteries is much higher—rated at 10 to 15 years, depending on the application—and depends on the design. Pocket-plate batteries have a useful life of 1,000 charge/discharge cycles, and sintered batteries are capable of sustaining 3,500 cycles. As with all batteries, the cycle life of a NiCd battery can be degraded from a number of issues, such as depth of discharge (DOD), temperature, and fast charge/discharge cycling. Unlike LA batteries, NiCd batteries have a far greater cycle life expectancy at small DOD—at only 10% DOD, life cycle expectations reach 50,000 cycles. Nickel cadmium batteries are also more tolerant of higher temperature operations, able to withstand occasional operations at 120°F, making them far more tolerant for nonconditioned spaces for standby power. As with LA batteries, NiCd batteries can be charged at varying rates with accompanying effects on the battery and the ability to transfer the full energy storage capacity of the battery. Nickel cadmium batteries require a somewhat greater amount of float charge as they lose between 2% to 5% of their charge per month at room temperature (LA batteries lose 1%). Most of the discharge occurs just after charging, and self-discharge increases with temperature.[41]

Cost issues

For many reserve power applications, nickel cadmium batteries are one of the least expensive energy storage technology choices for most applications in the retail and some transmission applications. According to the *EPRI-DOE Handbook of Energy Storage*, a multi-MW, multifunctional nickel cadmium battery system has a system cost of $600 per kW, with a little over 60% from the battery module. As with other mature energy storage technologies, there is no real expectation for significant cost reductions over the near term from technology or manufacturing advancements.[42]

Although NiCd batteries have a slightly higher initial capital cost compared to LA batteries, their superior operational properties actually provide lower cost of ownership. Possessing a higher cycle life and a lower maintenance requirement, these batteries have proved to be a viable substitute for LA batteries for many applications. Similar battery monitoring is necessary, but it can be done remotely, and maintenance requirements are significantly lower, requiring only annual inspection and care even for actively cycled batteries. As nickel cadmium batteries are also more tolerant to higher temperatures, environmental management and control common for LA or other battery systems will not add any relative additional costs to the installation, and may even reduce or eliminate them in certain circumstances. Finally, cadmium is a toxic material, requiring close control throughout production, use, and recycling of the material.

Installations

Nickel-cadmium batteries have been used in a number of applications in the electric power industry as a replacement for LA batteries.

Example—Golden Valley Electricity Association, AK.[43] The Golden Valley Electricity Association (GVEA) of Fairbanks, Alaska, had a problem. GVEA's transmission system is radial in nature, leading to power quality difficulties for customers when there are problems because of swings in the load, environmental forces, or a problem with the *intertie* (inter-tie) or power facility in Anchorage. After evaluating a number of options, GVEA installed in August 2003 the first four strings of a $35-million, six-string battery energy storage system using Saft's NiCd battery, which provides 40-MW, 10-MWh (15 minutes) (power electronics from ABB). When completed, this facility will be the largest battery system ever installed. The unit is designed to help improve the reliability of service to GVEA members and operates in several distinct modes, all of which help GVEA reduce its fuel costs. First, the unit provides frequency regulation by repetitively cycling from charging to discharging. This is a fundamentally different tactic from that of balancing a mismatched system load from existing generation facilities, which must constantly change their output level to provide this balancing role. Second, the unit reduces the high spinning reserve requirements common among Alaskan utilities because of their independent and divided nature. Finally, the unit is designed to cover the 15-minute period between loss of generation and the start-up of backup generation. To date, the facility has been a success; as of June 2004, 26 events totaling 336 minutes of activity were already encountered.

Example—Eskom, South Africa.[44] Eskom, the national power utility for South Africa, was having trouble with the backup LA batteries at its Lethabo power station for one of its generator units. These LA batteries provided backup power for a number of critical systems, which supported the local control systems and ensured a safe shutdown of the unit in case of the unit tripping off-line. These batteries were designed to have a lifetime of 20 years, but they were proving to only have a life span of 12 years. The cause for this was traced to the elevated temperature, which often reached 95°F for extended periods of time. Rather than simply replacing the batteries, Eskom decided to consider alternative battery technology and investigated a number of alternatives through a life-cycle cost evaluation. Through this analysis, Eskom decided to switch to Alcad NiCd rechargeable batteries for these critical control and protection systems. In addition to their ability to withstand the demanding temperature, the NiCd batteries also provided a number of other benefits over LA batteries. One of the biggest problems with the LA batteries was their unpredictable lifetime. They tend to have a nonlinear life and die suddenly; when their internal components corrode, their capabilities quickly deteriorate (hence, the need for constant monitoring). NiCd batteries, however, fail linearly and predictably; they have a much longer lifespan and require only infrequent maintenance. Since the NiCd batteries were installed in September 2000, they have proved successful, and plans are underway to replace the control backup systems for the other five units—and the backup power system for the entire plant's control system—with NiCd batteries.

Prospects and challenges

Nickel-cadmium batteries will remain an important energy storage technology in a number of existing market applications for the foreseeable future. Although slightly more expensive than LA batteries, their operating capabilities and excellent reliability span a wider operations envelope and longer cycle life, allowing for lower ownership costs. In particular, they suffer no sudden-death events common among older LA batteries.

Because of their low maintenance requirements, they continue to be a favorite for longer-term deployment in remote telecommunication and renewable energy applications. However, prospects for this technology to be used in the expanding role of energy storage technologies are limited because the requirements of many of these applications would significantly curtail the life of the unit.

Unfortunately, a number of challenges will continue to restrain significant further deployment of NiCd batteries for stationary applications in the electric utility market. For many standby power needs, customers still place significant emphasis on up-front costs rather than total ownership—although this is slowly changing. As these markets (UPS and remote power needs) continue to grow, NiCd batteries must continue to capture additional market share, although they will always remain more expensive than LA batteries for energy storage (but better for power delivery). Because of their improved operating capabilities, larger-scale NiCd-based battery energy storage facilities are somewhat more practical than LA systems, but not enough for widespread use in the wholesale power market. Although some advancement continues, some recent advances in the manufacturing process fall short of the breakthrough desired. As a mature technology, further development will continue, but much additional research into nickel-based battery technology to reduce manufacturing cost will continue to be directed toward other nickel-based battery technologies. As environmental concerns about toxic metals continue to grow, standards and regulations concerning the manufacturing, handling, and disposal of these batteries will continue to rise.

Major developers

There are a number of manufacturers of NiCd batteries for industrial and electric power market applications. Some of the larger manufacturers of NiCd battery systems include Alcad Limited, Hoppecke Batterien GmbH, Saft, and Tudor Batteries.

Flywheels

Summary

Flywheels have been an essential tool for leveling power flow in and out of a spinning mechanical device for a very long time. Flywheels store energy through accelerating a rotor up to a very high rate of speed and maintaining the energy in the system as kinetic energy (fig. 3–11). For this technology, there are two main veins of development, low- and high-speed rotors. Low-speed steel rotor flywheels predominate and are used primarily in uninterruptible power supply (UPS) devices. In high-speed flywheels, advanced composite materials are used for the rotor to lower its weight while allowing for the extremely high speeds. Although they could be used for UPS applications, the increased energy storage capability allows for other market applications, such as regenerative energy storage. For both types, energy is stored in the rotor in proportion to its momentum, but at the square of its surface speed—hence the desire to develop high-speed flywheels with correspondingly higher energy densities to allow for other market applications. The flywheel releases its energy by reversing the charging process and using the motor as a generator; as the flywheel releases its stored energy, the flywheel's rotor slows until it is fully discharged. Flywheels are suitable for applications requiring frequent and deep discharges that generally prove too taxing for standard battery installations and are also more compact and require less maintenance. The development of flywheel systems with far higher cycling capabilities supports a number of emerging applications such as regenerative energy applications.

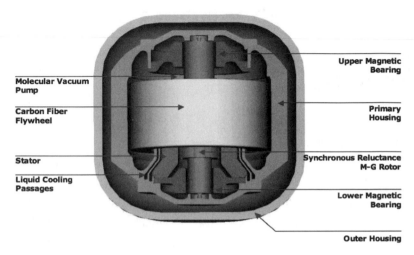

Molecular Vacuum Pump

Carbon Fiber Flywheel

Stator

Liquid Cooling Passages

Upper Magnetic Bearing

Primary Housing

Synchronous Reluctance M-G Rotor

Lower Magnetic Bearing

Outer Housing

Fig. 3–11. Flywheel (Courtesy of Pentadyne Power Corp.).

Historical origins

Flywheels have been used for thousands of years to both store energy and smooth out the variable-speed operation of rotating machines. Early hand-powered strategies such as a potter's wheel were expanded on during the industrial revolution to balance the variable power from such pulse sources as early steam engines. The flywheel, attached to the rotating shaft, moderates fluctuations in the shaft's speed by temporarily storing excess energy to be used during the nonpowered stroke of the engine; this more-manageable power source then, in turn, allowed for the development of far more complex powered mechanical drives. The subsequent use of steam and combustion engines to produce electricity introduced mechanically based flywheels in the power industry, with many still incorporated in small generators today. Development of flywheels as a stand-alone energy storage unit for electrical power was given support in the 1970s, when advances in power electronics allowed for the efficient voltage and frequency control of the power output regardless of the rotational rate of the flywheel. These electronics also allowed for a higher utilization of the flywheel's momentum, which, coupled with subsequent advances in material science (carbon fiber, magnetic bearings), allowed for greater energy densities and the potential for usable energy resources over a wider set of applications.

Design and operations

The basic components of a flywheel energy system include a rotor, motor/generator, bearing system, vacuum housing, and power electronics (conversion and system control). The rotor is the most important aspect of the system, because its design dictates the amount of energy able to be stored. Flywheels store power in direct relation to the mass of the rotor, but to the square of its rotational surface speed. Therefore, the best way to increase the amount of energy in a flywheel is to make it spin faster, not by making it heavier. However, it is important to understand that it is the surface speed that is important—not simply RPM—so a larger-diameter flywheel can have the same energy level as a smaller one that rotates much faster. Flywheels store energy through accelerating the rotor up to its operating speed and maintaining its rotational speed (and level of energy) with a small but constant additional energy input. When power is needed, the flywheel reverses this process and discharges its stored energy by using the motor as a generator; as the flywheel releases its stored energy, the flywheel's rotor slows. Because energy is stored mechanically and not chemically, it is possible to deeply discharge the unit repeatedly without any damage to the unit. To maintain the energy in the system, anything that provides resistance to the spinning rotor is minimized. Early bearing systems were made from roller bearings, but many designs today use either passive or active magnetic bearings to reduce frictional losses. To further reduce resistance, most high-energy designs also maintain the spinning rotor in a vacuum housing. Although necessary to increase the surface speed of the rotor, these last two design changes unfortunately also minimize the transfer of heat out of the unit (a constant concern to extend the life of the unit), so modern systems frequently have some type of cooling or chiller system included as well.

Because the power and energy components are decoupled in flywheels, these systems can be loosely classified into two categories, optimized for either power or energy. Whereas optimizing for power requires a greater emphasis on the motor/generator and power electronics, optimizing for higher energy densities requires a larger, high-speed rotor. Depending on the material used and design speed, the rotor's diameter can vary greatly. Low-speed flywheel systems generally have a heavy, solid steel rotor and rotate with speeds below 10,000 RPM. Having less energy, they are geared toward shorter bursts of power in such applications as power quality in a UPS. High-speed flywheel systems spin a lighter rotor at much higher speeds, potentially up to 100,000 RPM, but generally in the 20,000 to 60,000 RPM range. Because of the increased stresses at these speeds, high-speed flywheel rotors are normally constructed from composite materials, such as fiberglass or carbon fibers, impregnated in an epoxy and wound into a thick cylinder. Although high-speed flywheels can also be used for UPS applications, the increased energy storage capability allows for other, emerging market applications.

Flywheels used for energy storage have three basic applications: power quality, regenerative energy, and frequency control. First, flywheels have provided power quality enhancements as a substitute for batteries in UPS systems for many years now. Besides helping to ensure high quality, usable power service, they are able to provide a vital ride-through energy source until a backup generator can be brought online. Most notably, flywheel-based systems have proved to be far more capable in harsh environments where frequent and sometimes deep discharges would shorten the useable life of the common LA battery installation by 50% or more. Second, flywheels are capable of capturing wasted energy in repetitive motion systems by converting it to kinetic energy in the rotating mass of the flywheel rotor. Here, they are becoming increasingly popular in industrial crane or light-rail systems that undertake repetitive motions and frequently need short bursts or pulses of power. This not only protects against excessive wear and tear on the existing equipment, but also improves the economics of the application. Finally, the repetitive cycling capability of a flywheel is ideal to dampen frequency variation on power systems caused by moment-to-moment imbalances in power supply and demand. Here, support can be provided with a relatively small energy storage capability (as compared to the power grid), because the repeated charging and discharging can be accomplished quickly and continuously. Frequency regulation is of concern to all levels of the power grid but is of special concern on smaller power grids or isolated, self-generating industrial facilities. The variability of wind turbines is becoming another area where high-cycle storage facilities can be used to smooth the output of wind turbines.

As improved power electronics, vacuum housings, and magnetic bearings have become more widespread, round-trip efficiencies of flywheel systems have improved; many current production models are in the 70% to 80% range, with some newer designs even higher. Because these are mechanically based systems, their charge/discharge ratio is 1:1 and is capable of cycling tens of thousands of times. However, as the components of the system can be optimized for either power or energy, the needs of one application may make that design poorly suited for the other.

Cost issues

According to the *EPRI-DOE Handbook of Energy Storage*, initial capital costs of flywheel energy storage systems are in the range of $459 per kW. Although flywheel installations generally cost at least 50% more than LA batteries for similar (UPS) applications, their total life-cycle costs can be vastly less expensive, even while providing better reliability. Driving this competitive life-cycle comparison are the much lower operation and maintenance costs, plus much greater life of the flywheel versus the battery strings. These comparisons usually assume an average power quality environment; however, flywheels can provide the same level of capability in environments that would considerably shorten a lead-acid battery's life. Harsh temperatures and constant fast and deep charge/discharge cycling dramatically increase the cost of choosing batteries for these installations, but rarely affect the life of a flywheel. For regenerative energy and frequency response strategies, batteries are not an option, so flywheels must compete against standard business practices.[45]

Installations

Flywheels used for electrical energy storage are well established in the UPS market, but recent high-speed flywheels are supporting a growing number of alternative applications.

Example—STMicroelectronics, Rousset, France.[46] Normal operations in the semiconductor industry are extremely susceptible to power disturbances and thus require high-quality power at all times. To reach zero shutdown from power disturbances, STMicroelectronics installed, in 2001, a series of flywheel UPS systems from Active Power in two areas of its semiconductor fabrication facility in Rousset, France. The first application was to provide voltage support for variable speed drives (VSDs), and the second provided a ride-through for wafer fad production equipment. In the first placement, three 160-kW single flywheels were installed (1 per VSD) to keep the DC bus above the cutoff for the drives, as sags more than 22% above 10 ms were shown to turn off the VSDs. Within the first year, these units experienced 200 discharges, some of which would have resulted in shutdown of the drives. On the second application, five 600-kVA dual-wheel flywheel UPS systems were chosen to support the utility power for the wafer fabrication line. In both locations, flywheels were chosen because of their reliability, small footprint, and tolerance for temperature variations. Specifically, flywheels were chosen over batteries because the potential for a battery to fail within the 40-battery string was found to be quite high over the life of the unit, leading to the greater probability of a battery UPS failing in this application (10 times higher than flywheels). In addition, the extra room required (for extra batteries) to achieve the goal of zero shutdowns was not available because interior space is at a premium in a chip fabrication facility.

Example—Deluxe Films, Toronto, Canada. Precision batch processing of time-sensitive materials also relies on high-quality power. Here, a motion picture processing facility in Toronto, Ontario, Canada, had been feeling the impact of Toronto Hydro's inability to maintain high-quality reliable power during peak periods because of high local load growth. The processing facility, owned by Deluxe Film Laboratories, the world's largest motion picture film-processing laboratory, required high-quality power to operate the film-processing equipment. Over the last few years, the facility experienced numerous power spikes, sags, and even outages. Although many of these were short-term voltage problems, predominately lasting for five cycles or less, the company learned the true value of reliable, high-quality power during the 2003 blackout, when the facility lost power unexpectedly. Losing power like this—or even experiencing large voltage sags and spikes—created an enormous problem for the film company because it only takes one interruption to cause the equipment to shut down and lose the batch of production. If such a shutdown occurs during development of a critical piece of a large movie studio's production, the whole process must start all over. Not only does this produce a direct cost of lost material (multiple copies of the film) that can reach into the millions of dollars, future business may be lost from demanding clients. For these reasons, the company needed a means to ensure that the film production process could remain operational even during power outages or service interruptions. Therefore, in the fall of 2003 a SatCon Power Systems' STARSINE 2.2-MW rotary flywheel UPS system was installed at the facility. During installation, the unit was functionally demonstrated by running the entire facility during normal production without external utility system power to simulate a power outage similar to the one Toronto experienced during the 2003 blackout. Since installation, more than 25 power quality events sufficient to disrupt production have been recorded. The rotary UPS was able to mitigate each of them and prevent a shutdown or process failure, sometimes without the facility operators knowing—as was the case one day in April 2004 when the power system experienced so many problems that the unit engaged and remained in backup mode for more than four hours (from correspondence with Mike Gibson, Satcon Technologies of Boston).

Example—Lyon, France Metro.[47] The Lyon France Metro installed a 600-kW (4.3 kWh) Trackside Energy Management System from Urenco Power Technologies (UPT) in 2003 (3 x 200-kW units) to solve some chronic issues from overvoltage. During off-peak periods when not enough trains were on the network to absorb the excess energy generated by braking trains, the resulting overvoltage caused harmful wear and damage to the equipment onboard the rolling stock. Several other energy storage systems were evaluated but found to be not suitable because of the cyclical demands of the subway environment. The flywheel-based system acts as a short-term sink for the energy from the braking trains, with the regenerative energy used to power auxiliary station load, such as the lighting, ventilation, and escalators at the stop near the Hôtel de Ville. This solution has reduced the wear and tear to the equipment onboard the rolling stock and the brake maintenance costs.

Example—Usibelli Coal Mine, Usibelli, AK. Although most flywheels have relatively little energy storage capacity because of their size, some flywheels can actually be quite large. One such example was installed at the Usibelli Coal Mine in Usibelli, Alaska, for $1 million in 1982 and is used to support a dragline (a very large bucket-loader) at the coal mine. The electrical load from the dragline is an erratic 8-MW band (plus or minus 4 MW) every 60 seconds, necessitating some means to lesson the impact on the relatively lightly loaded GVEA power grid.

The flywheel itself is made from three one-foot thick, eight-foot diameter steel disks (total of 40 tons), with a normal operational speed from 900 RPM to 1,100 RPM. During operation of the dragline, the flywheel trades power with the dragline by slowing down during dragline loading and speeding up during dragline unloading. To support this activity, the flywheel can produce up to 5.2 MW for three-second (260 kWh) bursts, with any additional power required coming from the 1.8-MW motor/generator. With this flywheel system, the average power demand from the utility is around 2 to 2.5 MW, and GVEA only sees around a 500-kW band of power fluctuation (from personal correspondence with Tim O'Neil, Usibelli Coal Mine Inc.).

Prospects and challenges

Low lifetime costs and the ability to survive in harsh operating environments are the core strengths for the future success of flywheel energy storage technologies. Rapidly maturing from an already cost-effective base, life-cycle costs are already becoming a greater concern to owners—especially those who have already purchased earlier battery solutions for harsh applications. Flywheels currently represent 20% of the $1-billion market for the energy storage component of high-powered UPS market.[48] With the greater understanding of space conditions of other technologies, the minimal space and space-conditioning require-ments provide flywheel energy storage technologies a significant advan-tage in lifetime costs and usability—leading to a growing penetration of this existing market.

However, besides simply cannibalizing the battery share of the market, the capabilities of flywheels can also extend into applications not currently served by any solution because of the limited cycle life of many storage technologies. These markets—regenerative energy, frequency regulation, and the like—hold out the greatest opportunity in the long run for a much wider integration of flywheel systems into larger industrial settings. Coupling a flywheel with other power systems in stressful environments can extend the life of these other components by providing rapid injections/withdrawals to handle deep, cyclic events.

The main challenges toward furthering the introduction of flywheel technology are costs and customer education. Continued development of the technology will lower the initial costs of these units, but they will always have a higher initial cost than the LA battery solution. Therefore, increasing the understanding of the importance of life-cycle costs versus simply initial costs will be important for flywheels to increase their market share of the existing UPS market. Current pilot projects using flywheels for regenerative energy solutions or frequency regulation are also laying the groundwork for penetration into these other fragmented markets over the coming years.

Major developers

Current manufacturing of flywheel units is dominated by traditional steel rotor flywheel-based firms such as Active Power, Piller Inc. (rotary UPS and flywheel), Vycon Energy, Hitec Power Protection (rotary UPS), and SatCon Power Technologies (rotary UPS).

A number of other companies have begun developing higher-speed composite rotor based systems such as Pentadyne Power, Beacon Power, Boeing, and AFS Trinity.

Electrochemical Capacitors

Summary

Electrochemical capacitors are similar to batteries in that they have two electrodes immersed in an electrolyte and separated by a porous separator. The goal in this design is to obtain the energy storage capacity of a battery with the operating characteristics of a capacitor. They store energy via electrostatic charges on opposite surfaces of the electric double layer, which is formed between each of the electrodes and the electrolyte ions (fig. 3–12). Ultracapacitors move electrical charges between solid-state materials rather than through a chemical reaction; therefore, they can be cycled tens of thousands of times more rapidly and are not affected by deep discharges as are chemical batteries. Because the total amount of capacitance in the unit is directly related to the surface area of the electrode, energy stored increases with the square of the applied voltage. Their wide-ranging capabilities and reliability make electrochemical capacitors increasing favorites for retail markets, and they are promising for use in solving some transmission system stability applications in small niche applications. The prospects for electrochemical capacitors are strong, partially because of their infrequent use in the today's electric power industry. Currently, these technologies have grown into a $100-million market supporting mobile and communication systems, with growing inroads into fuel cell, motor starting, and the transportation market where size, weight, performance, and maintenance costs are important. However, significant—but reachable—hurdles must be overcome before electrochemical capacitors become widespread in the electric power industry. As a developing technology, their operational capabilities must be improved, their costs lowered, and their reliability (both in manufacturing and operation) must be enhanced.

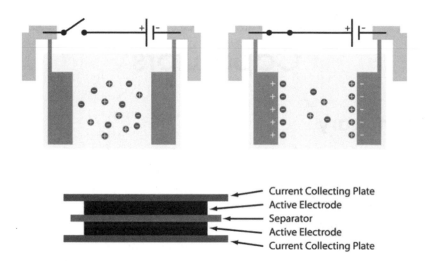

Fig. 3–12. Electrochemical capacitor (Courtesy of Maxwell Technologies, Inc.).

Historical origins

The first development toward electrochemical capacitors began more than 100 years ago, with some of the most promising research into a functional energy storage device occurring in the early 1960s at Standard Oil of Ohio (SOHIO). After the decision was made to not continue with the research program, SOHIO licensed all of its double-layer capacitor technology to NEC of Japan in 1971. After a few years of refining the technology and early manufacturing processes, NEC developed the first successful commercial electrochemical capacitor in the late 1970s for computer memory backup applications. Other companies continued to develop their own products and processing methods during this time. One was a Japanese electronics firm, Matsushita Electric Industrial Company, which patented a manufacturing method for improved electrochemical capacitor electrodes in the 1980s. By the 1990s electrochemical capacitors were scaled up and commercialized for pulse-power applications, fuel cell/engine starting applications, and specialty energy storage applications in the electric vehicle market. Electrochemical capacitors are often referred to by company trademark names including supercapacitors, ultracapacitors, and electric-double-layer capacitors.[49]

Design and operations

The goal of the electrochemical capacitor design—two electrodes immersed in an electrolyte and separated by a porous separator—is to obtain the energy storage capacity of a battery and the operating characteristics of a capacitor. It stores energy via electrostatic charges on opposite surfaces of the electric double layer, which is formed between each of the electrodes and the electrolyte ions. The electrodes are often made with porous carbon material, chosen because of the extremely large surface area available. There are four distinct design types of electrostatic capacitors, designated as types I–IV. Type I electrostatic capacitors are symmetric in design, with similarly sized positive and negative electrodes. The electrolyte is an aqueous solution with a high concentration of sulfuric acid. The maximum cell voltage is limited to 1.2 V, but it nominally operates at 0.9 V. Type II electrostatic capacitors, which are the most commonly used, are similar to type I, but with an organic electrolyte (ammonium salt in organic solvent). The maximum cell voltage is higher because of the electrolyte, with a maximum cell voltage of 2.3 V to 2.7 V. Type III electrostatic capacitors are asymmetrical in design. This design has a high cycle life partly because of the depth of discharge for one of the electrodes compared to the other. The electrolyte is an aqueous solution similar to type I electrostatic capacitors, giving a maximum cell voltage of 1.4 V to 1.6 V. Type IV electrostatic capacitors are new and not yet commercially available. They are asymmetrical in design, with an organic electrolyte similar to type II electrostatic capacitors, giving the design the potential for a maximum cell voltage of 4 V.[50]

Because ultracapacitors move electrical charges between solid-state materials rather than through a chemical reaction, they can be cycled tens of thousands of times, cycled more rapidly, and are not affected by deep discharges as are chemical batteries. Because the total amount of capacitance in the unit is directly related to the surface area of the electrode, the amount of energy stored increases with the square of the applied voltage. Because of the cell's charging properties, the voltage rises linearly with time when charged at a constant current, or when charged at a constant power. During charging, the electrolyte becomes polarized, with roughly half of the electrolyte material transferring an electron to the other half. These charged ions then migrate to the oppositely charged electrode, forming a charged layer on the surface—although no electrons are exchanged. These two layers of separated charges then form half of the electronic double layer, with a similar layer on the second electrode forming the other half. These are closely spaced, with the negative and positive charges separated by only half the diameter of the electrolyte ions.

The ability of electrochemical capacitors to respond quickly and reliably at full power is the key to their usefulness in the retail market, and they have the promise to solve some transmission system stability challenges. In the retail market, electrochemical capacitors are currently used for engine starting and pulse discharges used for power quality and bridging applications (UPS). For engine starting, large diesel generator and locomotive engines use them because they take up less space and weigh less than conventional batteries. They also start well in cold weather, have a long battery life, and are low maintenance. For UPS applications, electrochemical capacitors can either replace totally, or are currently being investigated for supplementing, conventional chemical batteries. By combining a front-end electrochemical capacitor with a chemical battery, the life of the batteries can be extended. In this setup, the electrochemical capacitors provide power for short duration interruptions and voltage sags, significantly reducing the cycling duty on the batteries so that they only provide power during longer interruptions.

As with all other energy storage technologies, environmental and operating conditions will have an impact on the scope and type of applications available to electrochemical capacitors. The average DC–DC round-trip efficiency of an electrochemical capacitor is 80% to 95% during normal operation, with variations because of the multiple design types. Temperature can affect the cycle life of an electrochemical capacitor. The normal upper operating temperature of an electrochemical capacitor is generally 85°C, with self-discharges increasing as the temperature rises; elevated temperatures decrease unit product life. Conversely, electrochemical capacitors have very good lower temperature operating characteristics, with the lower operating temperature for some variants being as low as −50°C. Because of their design and lack of moving parts or chemical reactions, electrochemical capacitors are capable of hundreds of thousands of charge/discharge cycles.[51]

Cost issues

According to the *DOE-EPRI Handbook of Energy Storage*, total system costs for an electrochemical capacitor are $456 per kW, with 40% of this attributable to the storage module. This cost cannot be readily compared to other energy storage facilities, as this is for a single-purpose facility rather than a multifunctional one supported by other energy storage technologies. However, for individual applications where space, weight, and lower maintenance requirements are key, this may prove decisive. Life-cycle capability becomes known as the unit ages, which is an important contributor to the cost-effectiveness of these versus LA batteries that can fail in sudden death—often requiring additional battery backups. As this is a recently commercialized product, significant additional cost reductions are envisioned, which will be primarily driven by manufacturing advancements and scale benefits from larger production runs rather than any new product breakthrough in material science. To date, automated manufacturing techniques have reduced the cost of electrochemical capacitors significantly. From the mid-1980s to today, manufacturers have reduced their costs 95%. With continuing manufacturing process improvements, it is estimated that the storage modules themselves can be reduced in cost by anywhere from 33% to 50%.[52]

Installations

Siemens' SITRAS SES System.[53] The braking of light rail systems remains an area of significant energy loss for city governments, and, therefore, significant effort has been made to reduce it. In the 1980s, many subway and light rail systems began replacing their friction wheel brakes with systems that reversed the train's electric motor and acted as a generator (for braking); then the regenerated energy was fed back into the rail power supply system. Unfortunately, much of this regenerated energy can only be used if there is a rise in demand proportional to a braking train—such as a train leaving a station; otherwise, only 66% or so of the energy will be used in the system. Without such an applicable load, the voltage of the system increases, so the remaining energy is dissipated as heat through radiators on the cars—raising the temperature in the tunnel and, thus, requiring greater power for air conditioning. To alleviate both the loss of energy and the voltage instability caused by this practice, Siemens developed the SITRAS Static Energy Storage (SES) system—a stationary electrochemical capacitor–based system rated at 1 MW, 2.3 kWh. An important deciding factor in choosing these electrochemical capacitors over batteries or other energy storage technologies is their long cycle life and low maintenance requirements. The unit is made from 42 Maxwell Technologies' 2,400 F cells (each with a capacity of 2,400 F). The SITRAS SES can automatically switch from «regenerative energy» to «voltage-stabilizing» mode, as needed. When the system is used on short-distance traffic—common on light rail systems—the power requirements at the station can be cut by approximately 30%. To date, trials are taking place in Portland, Oregon, and Dresden, Germany, with actual orders for systems received from Bochum and Cologne, Germany, and Madrid, Spain.

Prospects and challenges

The prospects for electrochemical capacitors are strong, in part because of the current lack of use of these technologies in the electric power industry. Currently, these technologies have grown into a $100-million market, supporting mobile and communication systems, with growing inroads into fuel cell, motor starting, and the transportation market where size, weight, performance, and maintenance costs are important. In particular, Russian companies continue to dominate this field for large electrochemical capacitors (needed for cold-weather engine and generator starting). Over the coming years, each of the different types of electrochemical capacitors is expected to improve the level of its energy density from 50% to 100%, potentially raising the energy density of some electrochemical capacitors to that of LA batteries—leading to even greater market penetration.

Significant—but reachable—hurdles remain to be overcome before electrochemical capacitors become widespread in the electric power industry. As a developing technology, their operational capabilities must be improved, their costs lowered, and their reliability (both in manufacturing and operation) enhanced. Core to all of these will be required progress along three avenues: capacitor design, material selection and usage, and manufacturing processing. Improvement along each of these avenues will lower the cost of each unit, and also improve reliability and the unit's ability to cycle heavily—a prerequisite for potential industrial and utility applications. For instance, large-scale systems with hundreds of electrochemical capacitors would require very uniform qualities of each cell (construction and operations) to prevent unbalanced electrical voltage and current flows over the life of the system. Although larger-scale manufacturing will provide significant cost-reduction opportunities, cost reductions through manufacturing process improvements will also be critical.

Major developers

There are a number of electrochemical capacitor manufacturers based in a number of countries, including Russia, Germany, Japan, France, Korea, and the United States. Some of the larger manufacturers and installers of these systems include ELIT, ESMA Joint Stock Company, NESS Capacitor Company, Maxwell Technologies, and Saft.

Superconducting Magnetic Energy Storage

Summary

Superconducting magnetic energy storage (SMES) systems store energy in the magnetic field created by the flow of direct current in a coil of cryogenically cooled, superconducting material (fig. 3–13). An SMES system includes a superconducting coil, a power conditioning system, a cryogenic refrigerator, and a cryostat/vacuum vessel to keep the coil at a low temperature—required to maintain the coil in a superconducting state and, thus, allow it to be highly efficient at storing electricity (more than 99%). These units can respond within a few milliseconds, and very high power output can be provided, but only for a brief period of time—therefore, these units are best suited to provide repeated, short-interval discharges. Although their costs are high compared to other storage technologies in respect to the cost per unit of energy stored, they are cost competitive with other flexible AC transmission systems (FACTS) equipment or transmission upgrade solutions, which are normally the competing choices. These facilities currently range in size up to 3 MW, and are generally used to provide power-grid stability in a distribution system and power quality at manufacturing facilities with critical loads highly susceptible to voltage instabilities.

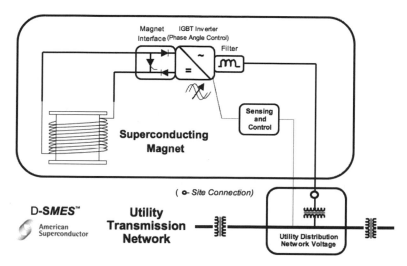

Fig. 3–13. Superconducting magnetic energy storage (SMES) (Courtesy of American Superconductor).

Historical origins

Based on the development of high-powered magnets in the early 1960s, the original concept for an SMES facility came in 1969 with a proposal for a large-scale unit able to provide commodity arbitraging capabilities—primarily diurnal opportunities associated with nuclear power. For the next two decades, a number of major programs for study of SMES began, centered on research in the United States, Japan, and Europe. Early U.S. research programs like the one begun at the University of Wisconsin in 1971 were supported and expanded on by groups such as the DOE, DOD, and EPRI. As mentioned earlier, much of this early focus and research on SMES was for large-scale energy storage, with the hope of storing hundreds of MWs or more as a competing technology for PHS facilities. However, the need for pulsed-power delivery for fusion power research led (among other expanding research opportunities) to further development of smaller SMES units for transmission system voltage stability and industrial power quality during the 1980s. One early installation in the United States occurred when Bonneville Power tested a 30-MJ SMES facility to provide area control and frequency regulation on long-distance power lines along the West Coast. Other installations included some in Japan, where developers such as Hitachi tested a variety of units to evaluate the system's capability of providing distribution line stability, one being a 5-MJ SMES unit at the Hitachi Works, and a second 1-MJ unit with Chubu Electric.

Design and operations

SMES systems store electrical energy in the magnetic field created by the flow of direct current in the coil of a cryogenically cooled superconducting material. An SMES system typically includes a superconducting coil, a cryogenic refrigerator, a cryostat/vacuum vessel to keep the coil at a low temperature, and PCS equipment. The superconducting coil can be developed in either a solenoid or torus shape. Although the torus design uses twice as much superconducting wire as the solenoid, its shape prevents the magnetic field from penetrating into the surrounding space, increasing efficiency. However, these same magnetic fields in the torus cause strong outward forces in the magnet and must be countered with strong reinforcements to prevent stress from degrading the magnet. For this reason, the solenoid design is used in current commercial designs.

Typically, the low-temperature superconducting coil is made of niobium-titanium, cooled to 4.2°K by liquid helium. The SMES system also uses high-temperature (77°K) superconducting components as an interface between the cryogenic area of the SMES and the conductors to the PCS. With the use of low-temperature superconducting material, the loss of the magnetic energy from resistance (heat) in the wire is avoided, enabling the superconducting coil to carry a very large current with little loss of power almost indefinitely—if the superconducting material is kept cold. By maintaining such a state, the energy storage capacity of the unit is enhanced because energy stored within the coil is proportional to the square of the current, and the coils vary in size depending on the energy storage capacity. The material properties of these superconducting coils are obviously of great importance, because temperature, magnetic field, and current density are all variables that trade off against each other in their design. In an ideal case, the material would maximize all three, but the physical properties and the manufacturing process normally degrade the performance of the coil from an ideal balance. During operation of the SMES unit (once charged), a small trickle charge is required to replace the power lost in the non-superconducting part of the circuit.

Although applications for just rapid power delivery are limited, coupling an SMES unit with sufficient power electronics gives the capability to rapidly inject both real and reactive power, providing relief in two markets: power transmission and the industrial sector. In the first market application—transmission voltage support—the SMES/power electronics unit provides fast response to voltage fluctuations. This protects the power grid from the destabilizing effects of short-term events such as voltage dips caused by lightning strikes and downed poles, sudden changes in customer demand levels, and switching operations. When the system detects a disruptive event (voltage drop) on the connected transmission line, these systems are able to respond within a few milliseconds, and in most cases can restore voltage stability to the power grid within one to two seconds (far less than the five seconds used as a guideline by many utilities). For these voltage stability roles, the first second is the most crucial, making response time a more valuable capability than gross energy storage capacity.

However, besides providing reactive power, experience with SMES units on the Wisconsin Public Service system (see example following) has shown that providing some real power significantly improves system performance and increases the rate at which voltage stability is restored during a voltage instability event. Finally, by providing voltage stability in this way, SMES power electronics units are able to increase the power transfer capacity of a locally congested power grid, often by upward of 15% depending on the layout of the system.

A second market where SMES units can be used is in the industrial power quality market to protect manufacturing operations from momentary sags on critical loads to prevent motor stalling or shutdown. A typical installation can react to voltage sags within a few milliseconds and can protect critical processes at the facility from voltage sags up to 60% for times exceeding one second.

SMES units are highly efficient at storing electricity, leading to a round-trip efficiency of more than 99%. For units targeting the utility market and the industrial market, the charging time for a unit is generally on the order of 90 seconds, with a full discharge taking less than 1 second, giving the system roughly a 90:1 ratio for charge/discharge cycle time. The main difference in these time requirements is the need to keep the temperature of the equipment in a superconducting state; more heating occurs during the charging than in the discharging event, necessitating a longer charging period. However, because these systems are not designed to cycle rapidly, this timing requirement is not of much concern as the more valuable market capability is the short discharge time—charging can easily occur after the system has returned to a stable state. Because of the solid-state makeup of the equipment, most SMES units have a cycle life of thousands of charges/discharges without degradation to the magnet, giving a unit a design life of 20 plus years.

Cost issues

According to the *EPRI-DOE Handbook of Energy Storage*, a multi-megawatt, single-functional SMES system has a system cost of $509 per kW, with a little more than 60% from the battery module. However, these costs are difficult to compare to other storage technologies because of the scale and purpose of the SMES units. SMES costs also are difficult to compare to other storage technologies in respect to the cost per unit of energy stored, and their market applications are somewhat different as well, so a direct comparison with other energy storage technologies is not always valid. More importantly, SMES systems are cost competitive with other FACTS equipment or transmission upgrade solutions, which are normally the competing choices. For example, in the Wisconsin Public Service example (described in a following section), American Superconductor's D-SMES unit, which costs $4 million, was found to cost far less (with a faster installation time) when compared to other solutions, which would have cost anywhere from $6 million to $15 million. Besides these up-front costs, operational expenses for these units are also comparable to, or less than, alternative technology solutions. With advancements in superconducting material capability and processing costs, the cost of the storage component has the potential to decline an additional 30%. Finally, the PCS equipment associated with SMES units is generally more sophisticated and is integrated as a larger component of the overall D-SMES system than in the other storage technology installations in order to handle the rapid discharge capability of the SMES unit and because the focus of the D-SMES' capabilities is in VAR production.[54]

Installations

SMES systems have been in use for several years at utility and industrial sites in the United States, Japan, Europe, and South Africa to provide both transmission voltage support and power quality to customers vulnerable to fluctuating power quality. In these two markets, more than 100 MW of these units (with the average unit being 3 MW or less) are estimated to be currently in operation around the world.

Example—Wisconsin Public Service.[55] In 2000, Wisconsin Public Service Corporation (WPS) installed six of American Superconductor's D-SMES units to handle voltage disturbances on the WPS Northern Loop system. Inductive motors powering the region's numerous paper mills comprise a large portion of the system's peak load, which is added to during the summer months by the significant air conditioning load of tourists in the region. Inductive motors pose special stability challenges to utility networks, as induction motor stalling at paper mills in northern Wisconsin sometimes precipitates the voltage collapse of the utility transmission system serving almost 33% of the state. To solve this instability problem, WPS has planned a transmission line upgrade but requires an interim solution until the upgrade can be moved through the planning process and brought into service. After evaluating other options (smaller power line extension, Static VAR Compensators [SVC], etc.) the D-SMES unit was chosen as having the least cost and lowest installation time. Once installed, the D-SMES units were able to solve the existing voltage instability problem while increasing power grid capacity by 15%. The installation has proved to be so successful that a seventh unit has been moved to the area to continue deferring the need for the additional transmission investment as the construction plans for the upgrade continue.

Example—Stanger, South Africa.[56] SMES units can also provide significant power quality benefits in the industrial market outside the oft-mentioned microelectronics industry. For example, a 1-MVA (3.0 MJ storage) Power Quality Industrial Voltage Regulator (PQ IVR) (with SMES unit) from American Superconductor (through Eskom, the local utility and partner) was installed at a Sappi paper mill in Stanger, South Africa, in 1997 to protect the plant from voltage sags (a voltage sag of only 250 milliseconds can shut down a paper machine). In this facility, a continuous paper web runs through several independently driven, speed-synchronized units. Any interruption in a drive snaps the web, resulting in downtime (measured in hours) for cleaning, rethreading, possible repair of damaged equipment, and finally restarting. These disruptions are significant, with production losses and damages at a typical paper mill averaging $50,000 per sag. The SMES unit has provided significant protection to the facility, protecting the mill from more than 70 voltage sags during its first year of operation—one-half of which could have caused the plant to shut down.

Prospects and challenges

SMES technology has found limited success to date, but real prospects for this technology exist along with opportunities for greater market acceptance and use going forward. The largest potential market available for SMES technology currently is to support utility transmission voltage levels against sudden disruptions. Here, two trends point toward growing market success for these units. First, current SMES systems compete well on price and capability against existing FACTS solutions and transmission upgrades for areas with growing transmission instabilities. Second, as additional investment capital is directed toward bolstering the existing but strained transmission infrastructure, SMES technologies stand to capture additional sales as this multibillion-dollar market expands. With the increasing constraints imposed on the transmission system from a growing load and scant system expansion, tools to provide system stability and increase the available transfer capability of the power grid are of increasing importance to utility power grid managers. As Charles Stankiewicz, vice president and general manager of American Superconductor notes, "Transfer of power that is limited by voltage stability can be effectively increased by solutions anchored by dynamic devices such as our D-SMES system. The end result is a more reliable network with an ability to import more economical electricity from outside the service territory."

The second target market for SMES technologies, industrial power quality, also has significant possibilities for added sales because high-speed and precision equipment continues to make inroads into manufacturing processes, with these machines being particularly inflexible to voltage instabilities. If an increased use of high-temperature superconductors becomes possible in the construction of these units, additional cost reductions from a vastly simpler and less expensive cooling system could also assist the competitive position of these units.

However, for all of these prospects, significant challenges still remain because SMES technology currently possesses only a limited installation base, leading to a greater level of misunderstanding of its actual installation requirements and operational capabilities. Although technologies competing to provide power grid voltage support and alleviate congestion also continue to meet a general reluctance, there is greater emphasis toward using these more traditional concepts. Essentially, the greatest issue of concern by utilities is the long-term reliability of the SMES unit in a utility setting rather than any cost or efficiency level of the unit. As is well understood, most utilities are skeptical of such technologies until their reliability and maintenance profile is proved elsewhere to be similar to comparable equipment. Solving this problem of customer understanding will only be possible through continued success and promotion of installations such as the ones at WPS and Entergy.

Major developers

The only major manufacturer of SMES products is American Superconductor of Westborough, Massachusetts. The firm sells its SMES product in cooperation with General Electric to utilities under the D-SMES line and to industrial customers under the PQ IVR name. The D-SMES unit is housed in a 50-foot truck trailer for easy installation at utility substations. As a mobile unit, the system provides transmission and distribution support that can be deployed as needed to reduce utilities' changing transmission constraints on their existing transmission and distribution power grids.

Thermal Energy Storage

Summary

Not generally thought of when discussing energy storage technologies, thermal energy storage (TES) systems are already well established as a means to reduce peak-cooling loads for commercial and industrial firms. Actively developed since the early 1980s, TES systems have advanced steadily so that now they are easily manufactured and installed as modular units, with nearly 7,000 systems active around the world that displace nearly 5 GW of peak load requirements. TES units are designed to work with the existing building's cooling system (a chiller), which chills either water or an ethylene glycol solution for the heat-exchange air conditioning of the building. The TES system simply uses the chiller to make ice or chilled water during off-peak hours and stores it in the insulated storage tanks (fig. 3–14). During the day, TES systems supplement the chillers by providing the cooling load for the commercial buildings, allowing for smaller chillers and substantially lowering air conditioning operating costs—both the amount of on-peak energy used and peak demand charges. On average, a retrofit installation has a one-to-three-year payback, and units integrated into new construction can often pay for themselves by reducing the amount of cooling system required for the facility.

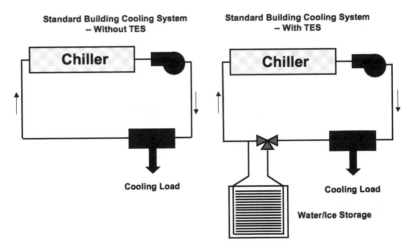

Fig. 3–14. *Thermal energy storage (TES) (Ardour Capital Investments).*

Historical origins

The development and use of TES—often referred to as *off-peak cooling*—for medium and large commercial cooling systems began in the late 1970s and early 1980s, when utility-sponsored conservation and load-sifting strategies were encouraged to avoid the need for new power plants as the cost of these assets was rising quickly. As the number of utilities wanting more load-shifting grew, state regulatory commissions enacted demand side management (DSM) programs to let utilities offer incentives for additional TES installations. By the early 1990s, many new manufacturers were producing TES systems for the commercial market. Unfortunately, many of these products were developed only to obtain the DSM rebate, without a clear ability to operate in the long term. This left many early adopters of the technology skeptical and left a misunderstanding of the technology in general. The early 1990s saw continued growth in the market, and realignment and maturing of the rebate strategies led to a decline in the weaker product designs. Since then, continued work toward improved coolants and validating the long-term economic savings have improved the standing of the technology. Although the move toward retail market deregulation in the mid-1990s initially hurt the industry with promises of cheap electricity and the loss of rebates, the reality of state-level retail market deregulation has shown that the basic economics for supporting TES—both for the consumer and the utility—are still valid.

Design and operations

A TES unit generally consists of a heat exchanger system of helical coils (metal or plastic) placed inside an insulated storage tank, a refrigerant pump, and an air-cooled condensing unit. Being well insulated, these storage tanks can normally be located anywhere, either indoors or outdoors. TES units are designed to work with the existing building's cooling system (a chiller), which chills either water or an ethylene glycol solution for the heat-exchange air conditioning of the building. The TES system simply uses the chiller to make ice or chilled water during the night and stores it in the insulated storage tanks. Besides being available because building cooling is not needed at night, chillers also perform more efficiently at night when the outdoor temperatures are relatively low (improving efficiency). When cooling is needed during the day, the TES system can then either assist or replace the chiller for meeting the day-time cooling load of the facility. If designed into new construction from the outset, the TES provides additional flexibility in the sizing of the entire cooling system needed. Normally, a building's chiller system operates to produce cold air for the building during the hottest part of the day, which happens to be during the peak daily load for the utility. For sizing purposes, these chillers are normally designed to be run on the hottest possible day, all day, at full load. For most usage needs, therefore, the unit is far oversized, so incorporating a TES system allows the building's chiller system to be sized and run more efficiently.

For these reasons, a TES system operates in an energy management role in the retail energy market for commercial users with medium to large cooling loads. The TES then essentially shifts building cooling from expensive on-peak electricity periods to off-peak periods. One important determinant of the true value of TES is the difference in peak versus off-peak utility rates. Overall, by operating in this manner, TES systems not only reduce the peak energy used, but also its cost. First, off-peak power is cheaper because of time-of-use rates for large energy users; and second, reducing peak demand allows for a reduction of the demand charge.

As TES systems are truly system optimizers for existing building chiller systems and not stand-alone units connected directly to the electrical load, comparing their efficiency directly to other energy storage technologies is difficult. On average, TES systems lose through thermal loss only approximately 1% of their energy during inactive use. This efficiency is comparable to or much higher than other technologies. Another way to look at the effective efficiency improvement offered by TES systems is the effect of shifting the load to off-peak periods. For instance, an earlier California Energy Commission (CEC) report explained that the summer peak heat-rate for an average steam unit in California was 11,744 Btu per kWh, compared to its off-peak idling heat rate of 7,900 Btu per kWh.[57] Using TES systems effectively arbitraged that difference for the utility and reduced the delivered price for the consumer. Generally the units operate in charging mode (making ice) for the same length of time as they do in discharging (supporting cooling load).

Cost issues

Typical costs for TES facilities can vary greatly depending on whether a unit is installed as a retrofit or as part of new building construction. For retrofit installations, exact costs are difficult to establish as each cost is project specific, but generally they range from $250 per peak kW shifted to $500 per peak kW shifted.[58] TES systems can be justified as a retrofit for customers with existing air conditioning equipment based on the lower costs incurred by using off-peak instead of on-peak power. Typically, however, the decision to proceed with a retrofit installation is not simply due to the cost savings in the energy bill, but also some other capital asset decision (or a combination), such as an aging chiller plant, building expansion, or limited electrical supply. Typical payback periods for these retrofit TES systems normally range anywhere from one to three years. This is possible through savings from reducing the demand charge and lowering peak power purchases. The demand charge in particular is a target for reduction because for a typical commercial facility, the demand charge can often equal 50% of the entire cost of service. With lower on-peak use of power, these facilities can reduce peak power demand by 50% and can reliably reduce cooling load costs up to 30% compared to standard cooling equipment.[59]

Integrating a TES unit with new construction (medium to large construction) provides for even greater benefits—in most cases, the decreased design requirements on the building's cooling system will pay for the cost of the system, plus provide the additional operating cost reduction described in the retrofit installation. In particular, because TES systems allow a building's air conditioning system to operate at lower temperatures, duct sizes can be reduced by 20% to 40%, and chillers can be 40% to 60% smaller than normally required.[60] Because the cooler air also requires a smaller volume of air to cool the building, fan motors, air handlers, and chilled water pumps are smaller and less costly. Based on the reductions in capital equipment needed when a TES system is integrated into the original design, EPRI has estimated that overall ventilation and air conditioning costs are reduced by 20% to 60%.[61]

Installations

TES systems have one of the widest application bases of all consumer-side energy storage systems, with roughly 7,000 units installed worldwide and a combined capability of shifting nearly 5 GW of peak-demand load. Although most of these units are for individual facilities, some are for multifacility systems and can become quite large. For example, in Chicago, Illinois, Unicom Thermal Technologies has installed a central 66,000-ton cooling unit that serves its commercial customers in the downtown area. (A *ton* is the standard metric for cooling load requirement—12,000 Btu heat removal per hour—and represents the amount of cooling energy melting one ton of ice over a 24-hour period.)

Example—Grossmont Hospital, La Mesa, CA.[62] Hospitals have large cooling loads, and Grossmont Hospital in La Mesa, California, is no exception. In looking for way to lower the cooling load cost of the 490,000-square-foot medical facility, the management of the facility became interested in incorporating off-peak cooling strategies. A feasibility study to determine the cooling load of the hospital found that the facility had a cooling requirement of 740 tons. It was found that by using a TES facility made up of 22 Calmac ICE BANK tanks, the chiller required could be reduced in size to a 320-ton unit. Besides allowing for a smaller chiller, the use of the TES unit qualified the hospital for a significant up-front subsidy from San Diego Gas and Electric to defray a portion of the system cost—which helped convince hospital officials to proceed with the retrofit project. The installation of the ice-storage system enabled Grossmont Hospital to shift 232 kW of load to off-peak and avoid paying a demand charge. Once operating, the cooling cost savings from the unit approached $17,000 per year, providing a payback period of approximately three years. Using the modular Calmac

ICE BANK system had other benefits; the unit could be easily expanded to handle additional cooling loads from planned additions to the hospital complex, such as a 60,000-square-foot Woman's Center complex that was slated to be added the following year.

Example—San Jacinto College District, TX.[63] TES projects are also important to state governments for the savings they promote—especially for the state-run colleges and schools. In this example, the Texas State Energy Conservation Office (SECO) helped the San Jacinto College District in East Harris County, Texas reduce its cooling load expenses by more than $1.5 million over seven years through the installation of a chilled-water TES unit at each of the three facilities. The original plan was devised in 1991, with the first unit installed in 1994 and the other two units following soon thereafter. The three TES systems at San Jacinto generate a total of 1.7 million gallons of chilled water during off-peak electric-use hours which is then stored until it is needed to cool facilities during peak hours. The $2.7-million loan for the entire project was financed through SECO's Texas LoanSTAR program, which allows borrowers to repay loans with money saved by the energy retrofits. Projects in this program usually are expected to produce enough savings to pay the loan back within eight years. Also benefiting from the reduction in peak demand, Reliant Energy gave the college a $500,000 rebate for installing the unit (for load repayment).

Example—Villa Julie College, Baltimore, MD.[64] The real value of a TES unit often is realized through its variety of benefits. In this example, Villa Julie College was adding more than 135,000 square feet of space to its campus in 2000, effectively doubling the size of the college, which is located outside of Baltimore, Maryland. One of the primary objectives of the new facility's $12-million design was energy efficiency. To support the decision for the inclusion of TES units, Baltimore Gas and Electric provided a rebate of $52,400 toward the units' installation. For these reasons, the college chose Baltimore AirCoil Company's Modular ICE CHILLER® thermal storage units to be part of the new installation.

In this system, two 300-ton chillers supply ethylene glycol at 19.2°F from 10:30 p.m. to 6:30 a.m. to build ice in the storage units. During the day, only one of the chillers and the ice storage system are then required to support the cooling load of the new facilities. Besides enhancing the energy efficiency of the new construction, the TES units were essential because the building's design called for a much-reduced internal space allotment for mechanical equipment. The TES units worked well with the smaller ductwork and piping and did not affect the internal aesthetics of the design. The results of this new system have been impressive. By shifting 262 kW of on-peak demand, the units have provided a $44,700 annual cost savings, and over the expected 20-year life of the units, this system is expected to provide more than $460,000 in value compared to a conventional cooling system. Plans also exist for further savings by integrating the two cooling systems on campus. Currently, two chillers meet the cooling demand of the existing buildings. When the two systems are finally connected, the cooling requirements of the campus will be met by the ice storage system and by operating only the two new chillers.

Prospects and challenges

TES is a mature technology that provides real potential saving for customers—especially those with large cooling loads—as either a retrofit or in new construction, because central chillers are used in approximately 80% of buildings greater than 200,000 square feet. Continued development will help improve the economics of the technology, but the real strength in the argument for using TES facilities is the growing body of evidence for successful installations of these units, which can more directly supplant existing hesitation toward these systems. Another important aspect for the continued deployment of this technology is the ongoing utility support because of the savings potential with the continued installation of these units. Shifting peak demand load has long proved to be less expensive than building new power facilities or ensuring power from independent generators is available. Striking closer to home for these utilities is the ability to slow or reduce peak demand on a particular distribution line and thus prevent the need of an upgrade for some time. Both of these benefits point toward utilities' continued support through small but important up-front grants to customers to defray the installation costs of these units. As these grants are also approved through state PUCs, continued government support is also important. For these PUCS, their continued support stems from not only the lower end-user costs these systems can provide, but also their support in reducing peak-power demand, which helps to lower the environmental impact required to meet the peak load.

Challenges for these systems will continue to be present, however, as their benefits are misunderstood to only be long-term and, as with any piece of capital equipment, they require some up-front capital outlays. Continued changes and uncertainty in the makeup of retail energy market rules and their resulting retail consumer price structure are also of concern—not for how they will affect the economics of the installations, but how they affect the perception of future retail prices. Finally, other building efficiency improvements such as improved lighting, windows, and the like will continue to compete for any available funding in the overall conservation strategies to reduce the cooling load costs.

Major developers

There are a number of manufacturers of TES systems based on varying technologies. Some of the larger manufacturers and installers of these systems include Baltimore AirCoil Company, Calmac Manufacturing Corporation, Cryogel, Dunham-Bush, Inc., FAFCO, Inc., Evapco, Henry Vogt Machine Company, and Paul Mueller Company.

References

1. Ogelthorpe Power. Rocky Mountain Pumped Storage Hydro Plant. http://www.opc.com/opccom/power/rocky.jsp?menu=power, http://www.opc.com/ (accessed May 1, 2005).

2. Dinorwig and Electric Mountain, First Hydro Company. http://www.fhc.co.uk/DIN.htm, http://www.fhc.co.uk/INTRO.htm (accessed May 1, 2005).

3. Japan Commission on Large Dams. Seawater pumped-storage power plant. http://www.jcold.or.jp/Eng/Seawater/Seawater (accessed May 1, 2005).

4. Callahan, T., J. Degnan, and D. Miller. 2003. Upgrading the Taum Sauk pumped storage project. Paper presented at Waterpower XIII meeting, Buffalo, New York: HCI Publications.

5. Electric Power Research Institute (EPRI). 2003. Compressed air energy storage. *EPRI-DOE handbook of energy storage for transmission and distribution applications*, 15-1–15-42. Palo Alto, CA: Electric Power Research Institute.

6. Crotogino, F., K. Mohmeyer, and R. Scharf. 2001. Huntorf CAES: More than 20 years of successful operation. Paper presented at American Society of Mechanical Engineers, Spring 2001 Meeting, Orlando, FL.

7. van der Linden, S. 2002. CAES for today's market. Paper presented at Electrical Energy Storage Applications and Technologies meeting, San Francisco.

8. Iowa Association of Municipal Utilities. February 2003. *Transforming wind power into a reliable resource* (stand-alone 2-page handout).

9. CAES Development Company. 2002. Norton Energy Storage Project. (Presentation).

10. Electric Power Research Institute. 2003. Vanadium redox batteries. *EPRI-DOE Handbook*, 10-1–10-27; Lotspeich, C. 2002. A comparative assessment of flow battery technologies. Paper presented at the Electrical Energy Storage—Applications and Technology (EESAT) 2002 Conference, San Francisco, CA.

11. Electric Power Research Institute. 2003. Zinc bromine batteries. *EPRI-DOE handbook*, 9-1–9-26; Lotspeich, C. A comparative assessment of flow battery technologies.

12. Lotspeich, C. A comparative assessment of flow battery technologies; Electric Power Research Institute. 2003. Polysulfide bromide batteries. *EPRI-DOE handbook*, 11-1–11-25.

13. Electric Power Research Institute. 2003. Vanadium redox batteries. *EPRI-DOE handbook*, 10-1–10-27; 9-1– 9-26; 11-1–11-25.

14. Clarke, S. 2004. Observations on building a flow battery company. Paper presented at the 2004 Electrostatics Society of America (ESA) Conference, Columbus, OH.

15. Lotspeich. A comparative assessment of flow battery technologies.

16. Sumitomo Electric International. User: Tottori Sanyo Electric Co., Ltd. http://www.sei.co.jp/redox/e/index.html (accessed October 2004).

17. Blackaby, N. 2002. VRB technology comes to the fore. *Power Engineering International* 10, no. 3.

18. Williams, B. 2004. A case study on the use of VRB energy storage system (VRB-ESS) as a utility network planning alternative. Paper presented at the 14th Annual Electricity Storage Association Conference, Columbus, OH.

19. ZBB Energy. Project status-United Energy 400 kWh system. http://www.zbbenergy.com/status.htm (accessed September 2004).

20. ZBB Energy. Project status-Detroit Edison/Department of Energy and SNL 400 kWh system. http://www.zbbenergy.com/status.htm (accessed September 2004).

21. ZBB Energy. Project status-Australian Inland Energy 500 kWh system. http://www.zbbenergy.com/status.htm (accessed September 2004).

22. Regenesys Technologies. Project status-Little Barford Power Station. http://www.regenesys.com (now offline) (accessed November 2003).

23. Regenesys Technologies. Project status-Tennessee Valley Authority. http://www.regenesys.com (now offline) (accessed November 2003).

24. Electric Power Research Institute. 2003. Sodium sulfur batteries. *EPRI-DOE handbook*, 8-1–8-31.

25. Ibid.

26. Ibid.

27. Hyogo, T. 2003. Commercial deployment of the NAS battery in Japan, Takayama, Tokyo. Paper presented at the 2003 EESAT Conference, San Francisco, CA.

28. Baba, Y. 2004. Electricity storage applications for electric utility business. Kyushu Electric Power Co., Inc., Paper presented at the 2004 ESA Annual Meeting, Columbus, OH.

29. Nichols, D., B. Tamyurek, and H. Vollkommer. 2003. Sodium sulfur battery (NAS) applications. Paper presented at IEEE Power Engineering Society meeting, Toronto, Ontario, Canada.

30. Electric Power Research Institute. 2003. Lead-acid batteries. *EPRI-DOE handbook*, 6-1–6-51.

31. *World lead-acid battery markets*. 2002. San Jose, CA: Frost & Sullivan.

32. Linden, D., and T. Reddy, eds. 2002. Lead-acid batteries. *Handbook of batteries*, 3rd ed., 23.1–23.88. New York: McGraw-Hill.

33. Electric Power Research Institute. 2003. Lead-acid batteries. *EPRI-DOE handbook*, 6-1–6-51.

34. Ibid.

35. Ibid.

36. Faber De Anda, M., and J. Boyes. 1999. *Lessons learned from the Puerto Rico battery energy storage system*. SAND1999–2232. Washington DC: U.S. Department of Energy.

37. Sandia National Laboratories. 2000. *Energy 100 awards: Metlakatla energy storage system*. Sandia, NM: Sandia National Laboratories.

38. *World lead-acid battery markets*. Frost & Sullivan.

39. Linden and Reddy. Industrial and aerospace nickel-cadmium batteries. *Handbook of batteries*, 26.1–26.29.

40. Electric Power Research Institute. 2003. Nickel-cadmium and other nickel electrode batteries. *EPRI-DOE handbook*, 7-1–7-40.

41. Ibid.

42. Ibid.

43. Golden Valley battery energy storage system comes online. September 2003. *Electricity Storage Association Newsletter.*

44. ALCAD. Lethabo power station switches to nickel-cadmium batteries for back-up power for critical systems. http://www.alcad.com/about_ news_detail.asp?PressID=21 (accessed October 2004).

45. Electric Power Research Institute. 2003. Nickel-cadmium and other nickel electrode batteries. *EPRI-DOE handbook,* 7-1–7-40.

46. Active Power. 2001. *Flywheel energy storage for quality power in the semiconductor production industry.* brochure. Austin, TX.

47. Urenco Power Technologies. Case Study—Lyon Metro. http://www. uptenergy.com/eng/applications/traction/casestudy/lyon.htm (accessed May 2005).

48. *World UPS markets.* Frost & Sullivan.

49. O'Brien, D. 2001. EC capacitors deliver high capacitance in a small size. *Power Electronics Technology* 27, no. 3.

50. Electric Power Research Institute. 2003. Electrochemical capacitors. *EPRI-DOE handbook,* 14-1–14-46.

51. Ibid.

52. Ibid.

53. Schneuwly, A., J. Auer, and G. Sartorelli. June 2004. More power for the rails. *Power Systems Designs Europe* 1, no. 5: 26–27. AGS Media Group, Laboe, Germany.

54. Electric Power Research Institute. 2003. Superconducting magnetic energy storage. *EPRI-DOE handbook,* 12-1–12-3.

55. American Superconductor Case Study. 1999. *Wisconsin Public Service Corporation.* brochure. Middletown, WI.

56. American Superconductor Case Study. 1998. *Sappi Paper Mill, Stanger, South Africa.* brochure. Middletown, WI.

57. California Energy Commission. 1996. *Source energy and environmental impacts of thermal energy storage.* California Energy Commission, Report #500-95-005. http://www.energy.ca.gov/reports/reports_ 500.html (accessed October, 2004).

58. Thermal energy storage—Economics and benefits. 2002. E3 Energy Services, LLC., Arlington, VA.

59. Silvetti, B., and MacCracken. 1998. Thermal storage and deregulation. *ASHRAE Journal* 40 no. 4.

60. MacCracken, M. 2003. Thermal energy storage myths. *ASHRAE Journal* 45 (9):36–42.

61. Electric Power Research Institute. July 1991. *Cold air distribution with ice storage.* Brochure CU-2038. Palo Alto, CA, EPRI.

62. Calmac Manufacturing Corporation. November 2001. *Hospital market brochure* (Englewood, NJ).

63. Jones, D. January 2000. *Hoarding hertz.* Texas: Fiscal Notes, Texas Comptroller of Public Accounts, State of Texas, Austin, TX.

64. Baltimore AirCoil Company. 2000. *Ice thermal storage—PRJ42.* Baltimore, MD. http://www.baltimoreaircoil.com/english/products/ice/tsum/index.html (accessed May 2005).

4 APPLICATIONS

Energy storage technologies will have far-reaching impacts throughout the electrical power industry. They can conceivably provide benefits (fig. 4–1) in all three market segments: wholesale power, transmission and distribution, and the retail market. Within each market there are similar needs: to improve asset usage, increase flexibility and optionality, and prevent power variations from interrupting operations or damaging equipment. Storage technologies provide an extra dimension to make these improvements—time. Because they bring flexibility, energy storage technologies are useful in many market roles. They are best thought of as enabling technologies, either raising system optimization or promoting a market change, such as the faster introduction of renewable energy resources. Because they impact the market's margin when costs are highest and/or stress on the system is at a maximum, even a relatively modest action—especially a fast-acting one—could dramatically impact the market.

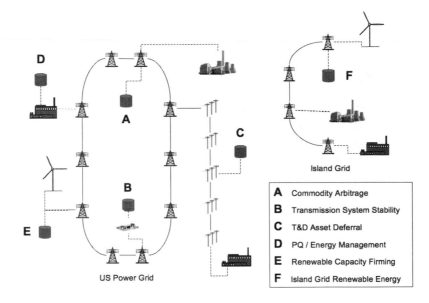

D

A

B

C

E

US Power Grid

Island Grid

F

A	Commodity Arbitrage
B	Transmission System Stability
C	T&D Asset Deferral
D	PQ / Energy Management
E	Renewable Capacity Firming
F	Island Grid Renewable Energy

Fig. 4–1. Energy storage applications (Ardour Capital Investments).

However, no energy storage technology is suitable for all market roles. Although most storage technologies could be used on innumerable applications, any energy storage technology is best suited for only a few related applications (fig. 4–2), where its technical capabilities can be leveraged for maximum economic benefit. This matching of the technology to the application is balanced between technology capability and application requirements. These roles are often grouped (fig. 4–3) according to how long the facility is expected to discharge power quality, bridging power, or energy management. Generally, short-duration storage roles are focused on damage- or loss-prevention, whereas long-duration storage roles are focused on commodity arbitrage. Because they can operate across a wide spectrum of uses, a technology can be optimized for one of the following applications above another by altering such attributes as the capacity of the storage medium, the conversion capability, and the like:

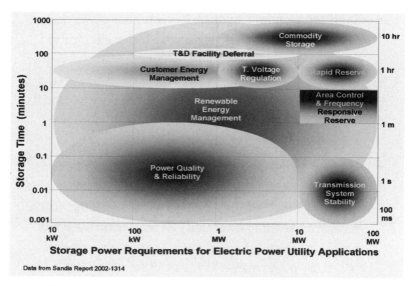

Fig. 4–2. Application capability sizing requirements (Courtesy of the Electricity Storage Association).

Fig. 4–3. Energy storage market roles (Courtesy of the Electricity Storage Association).

- **Short term—power quality:** Prevents poor power quality from damaging sensitive electronics or other electrical equipment from unpredictable and momentary changes (milliseconds to seconds) in the frequency, voltage, and so forth. Depending on the facility's size, these systems can provide voltage and frequency control (wholesale power/transmission market) or power quality (retail market).

- **Medium term—bridging power:** Assures a continuity of service (seconds or minutes in length) when switching from a primary generation source (utility service) to another (backup generator/alternate power feed). Depending on the size of the facility, these systems can act as contingency reserves (wholesale power market) or as a UPS (retail market).

- **Long term—energy management:** Allows arbitraging between two market periods (hours) by decoupling the timing of generation and electricity consumption, or *load leveling*. Depending on the facility size, these systems can be used for bulk power sales (commodity arbitrage), minimizing capital outlays (transmission asset deferral), or reducing individual firm energy costs (peak shaving).

Key Design Issues

As with any engineered system, the final design of every energy storage facility is essentially a compromise between attributes to optimize the facility for a particular application. There are three key issues in particular: energy versus power, cycling, and usage costs. Adjusting one part (i.e., greater power deliverability) frequently impacts the overall design and affects other attributes, most prominently the economics of providing the service for a particular application. For instance, power quality requires an extremely fast reaction time to be of any value to the consumer, with this requiring not only a high deliverability from the energy storage medium, but also from the supporting equipment, such as power electronics. Thus the importance of each capability is evaluated and balanced with the application in mind, along with the incurred cost of promoting one capability over the other. Other factors outside the direct operation of the unit can also have impacts on the design, even acting as a limiting factor, such as a physical footprint (limitations) or environment control requirements (space conditioning).

Energy versus power

The first aspect of an energy storage technology is the unit's capability to store and release energy at different power ratings. Outwardly, these aspects of the technology are responsible for determining the market role of the technology. For instance, the energy rating (kWh) is sometimes thought of as the *volume* or scale of the facility and is usually the prime determinant in how long a unit can operate. The power rating (kW) of an energy storage facility describes the rate at which it can absorb and discharge energy, and largely establishes in what part of the market the unit can operate—MW-sized units for the wholesale power market versus kW-scale units that are more suited for on-site installation at commercial or industrial facilities.

Inwardly, these application requirements for a specific size and capability will define both the type of technology capable of fulfilling the market role and the geometry of the storage facility. For instance, the ratio of energy to power is largely fixed for chemical batteries, whereas other technologies have separated these two, providing a degree of freedom in system design. Therefore, fitting a chemical battery to a particular market role requires adjusting the number and length of battery strings to get the requisite power and energy rating. For those technologies (such as flywheels or flow batteries) where these aspects are separate, fitting a technology to a market role entails scaling each aspect individually to what is needed. For instance, changing the power rating is normally accomplished by altering the conversion component (motor, cell stack, etc.) to improve the rate of power transfer of the storage medium. To increase the energy capacity, one would simply add (or enlarge) storage modules or increase the energy density of the storage medium. Although adding additional storage modules is generally the easier and cheaper option, outside issues such as space constraint can limit this choice.

Cycling issues

The second issue concerning the design of an energy storage technology is its cycling capability: the unit's ability to be repeatedly charged and discharged. All storage facilities have a finite useful life based on the number of times the unit is used; this plays directly on the applications they can support. This *cycle life* also varies depending on the physical attributes of the storage medium and how the energy is stored. For example, chemical batteries generally have a shorter cycle life (hundreds or a few thousands) than a mechanically based flywheel unit that can sustain hundreds of thousands of cycling events. This is the case as the chemical battery frequently has a small but cumulative chemical by-product during each cycling event, whereas the flywheel suffers little or no residual stress to the rotor. For these higher-cycle-life technologies, often the limiting factor is not the storage medium but its associated equipment. As these physical characteristics affect the cycle life of the technology, the cycle life itself then plays a central role in the usage cost of the system (fig. 4–4).

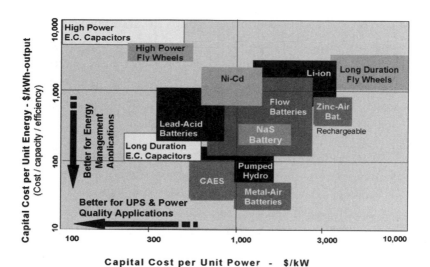

Figure 4–4. Cycling issues (Courtesy of the Electricity Storage Association).

Many factors affect the design cycle life of these technologies; three crucial factors are depth of discharge (DOD), the rate of discharge, and the environmental conditions in which the unit operates. Probably the greatest determining factor affecting the cycle life of storage technology is the DOD to which the unit is repeatedly cycled. Again, one can see differences between the technologies as some are electrical or mechanical (capacitors and flywheels) and are able to discharge almost completely without much effect, whereas the chemically based (lead-acid batteries) suffer significant deteriorations in their cycle life from full discharges. Below an 80% DOD, the cycle life of many of these chemical-battery systems deteriorates dramatically; for this reason, most comparisons between technologies are based on an 80% DOD. Second, forced, rapid recharging can also damage the storage medium and lower the cycle life of the unit or even its ability to charge. This is of greater concern to many of the chemically based storage technologies but can easily be avoided by limiting the rate of flow through the accompanying power electronics. Finally, environmental conditions can also negatively affect a storage medium's cycle life, especially chemically based technologies. These impacts also can be avoided or greatly reduced through the use of space conditioning, but this adds to the parasitic load of the facility, lowering the unit's round-trip storage efficiency.

The efficiency in each charge/discharge cycle, the *round-trip efficiency*, is also an important aspect of the cycling operation of the unit. These losses stem from resistance, heat, and so forth, as well as in the associated power electronics (although some comparisons of the technologies do not include this component). This round-trip efficiency determines to a large degree with what type of market application the facility can compete. Low round-trip efficiencies effectively increase the average cost of operating the unit, for example, so such a technology would be better suited for a power-quality role above commodity arbitrage—because the value of power quality is to prevent a catastrophic loss, whereas arbitraging power is based on the marginal difference in the price of a given amount of power. Other losses (float) occur while the unit is charged but inactive, and play a role in the economics of the unit to provide a competitive service for an application requiring long-term storage of energy.

Ownership costs

Although technology capabilities are important, for most potential owners the cost is the real deciding factor. At the time of purchase, this would consist of the up-front capital costs, but the total ownership cost should actually be used for any decision-making purposes. Different technologies have different operational life spans and operational economics that will greatly affect this calculation (fig. 4–5). The structure of these costs is important, especially when comparing different storage technologies. For some technologies the operational cost can end up an appreciable part of the total ownership costs. Although some technologies may be cheaper initially, the requirements of the application make the real usage cost far higher than a competing one, such as the technology's cycle life or round-trip efficiency. If purchasing a UPS unit, for example, simple lead-acid batteries often have the least initial cost. If the total ownership costs of the unit are evaluated, however, especially in a hostile environment (concerning power quality), flywheel systems are frequently the cheaper option.

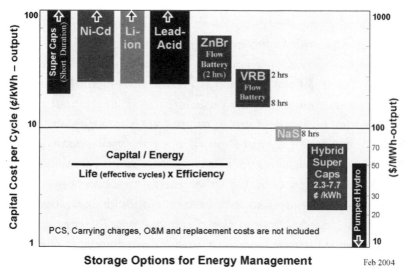

Fig. 4–5. Ownership costs (Courtesy of the Electricity Storage Association).

Capital and operating costs of a technology rely heavily on the design of the technology. Capital costs are based on many factors, including the technology's complexity and the scale of the facility; for this reason, storage technology capital costs are evaluated in dollars per kW like other power technologies. Depending on the technology in question, these costs can further be broken down into three subsets: costs that scale with energy (storage capacity of the unit), costs that scale with power (conversion module), and the balance of plant costs (system housing, monitoring system, etc.) that often only vary slightly with the size of the unit. Operating costs, normally measured in dollars per kWh (or dollars per kWh per cycle), originate from a number of sources, including required maintenance, consumables (for some chemical storage mediums), and replacement part costs. Included in these costs is the energy needed to maintain the unit in a ready but inactive state, and the parasitic load from supporting systems such as space conditioning. All storage mediums slowly lose slight amounts of energy—*float losses*. This is one aspect that determines whether some technologies are appropriate for long-term storage applications.

Choosing one technology over another should be done with care. This is especially true for finding a comparative metric, which holds biases. Any metric—such as leveled dollars per kW, dollars per kWh, and dollars per kWh cycles—holds limitation as a comparative metric for dissimilar technologies (pumped-hydro versus capacitors) in disparate market applications, yet is appropriate in comparing technologies with similar capabilities targeting the same applications. For instance, *dollars per kW* is an obvious guide for up-front costs. Similarly, *dollars per kWh* is useful to evaluate the cost of usable cycling energy, but *dollars per kWh per cycle costs* is more useful to compare technologies for applications with high-cycling needs, but not low-cycle applications. Therefore, the final choice should be made based on a variety of comparisons in relation to the application desired.

Wholesale Power

Energy storage technologies can play an important—and profitable—role in the wholesale power market both by supporting the efficient operation of existing power facilities and providing ancillary services to enhance system reliability and security. These facilities can also help promote the greater use of renewable resources (wind) and solve such issues as its unpredictable and noncoincident power production.

By decoupling electricity supply and demand, storage units can play a role in addressing many existing problems in the wholesale power market, such as the daily load curve, which creates two major problems for power facilities: chronically low usage and increased cycling stresses. As discussed previously, the current power facility fleet design and operation was crafted around the classic baseload, mid-merit, and peaking regime, which although it provides power in the most economical way, keeps overall usage of many fossil-fueled generation facility assets at 60% to 70%. Although peaking facilities may have been built to operate on a short-term basis, coal-fired units consistently operate below their optimal level. Raising the use of these units would not simply give them additional operating revenue. They would also operate more economically because the stress from cycling the units would be reduced, prolonging the replacement cycle of many hot-zone components of these boilers. Even combined-cycle, gas-fired facilities could benefit; many of the early units suffered far greater and faster cycling degradation than first envisioned. With storage units supplanting power production facilities in support roles, such as load following or contingency reserves, existing facilities can be redirected toward more stable power production.

This market is dominated by large power facilities; for that reason, only large-scale energy storage facilities measuring in the 10s, or more likely 100s, of MWs with discharge endurance of many hours will most likely be successful. Storage technologies like existing pumped-hydro facilities or even new compressed air energy storage (CAES) units can then be used to act as both a sink and source of energy in ever-changing market conditions. These larger units will cycle (inversely) on a daily basis with the system load or act to provide a ready reserve of power in the event of an incident. Smaller storage technologies can also operate successfully in these roles if the market they participate in is a smaller or isolated power grid on a much smaller scale. Like other assets in the power market, their capability is only one determinant of their role. Where the unit is situated will also impact greatly as to how and when the facility will best be able to leverage its capability.

By acting as both a sink and source of power, energy storage facilities can play a role in the wholesale power market along three broad applications:

1. **Commodity arbitrage:** Absorbs low-cost, off-peak power and resells it during peak demand periods when its value is highest. Besides earning revenue for the storage facility, this can provide benefits to both underused baseload units during off-peak periods and mid-merit units being damaged from excessive ramping to meet rapidly changing load requirements.

2. **Contingency reserves:** Provides additional resources in the event of an incident at a power or transmission facility. Besides providing these services directly to the independent system operator (ISO) in competition with generating facilities, these units could also represent a callable resource during peak periods to free up generation units held at standby.

3. **Blackstart capability:** Helps reenergize the power grid in the event of a blackout. By using centrally dispatchable and strategically placed storage units during these episodes, these facilities could hasten the restoration process.

Commodity arbitrage

Commodity arbitrage is the act of absorbing low-cost, off-peak power and selling it during peak demand periods when its value is highest (fig. 4–6). Facilities with the ability to dispatch large quantities of power on demand in a competitive market, such as pumped-hydro units, are already highly valued in the market as evidenced by the prices they were able to command during 1990s auctions. By definition, only storage facilities can provide this service because other power sector technologies cannot hold energy in a ready-reserve state. Because these units will operate in the wholesale market, the scale of these units would range from the tens to hundreds of MWs. Depending on the need, either the power from these storage facilities can be supplied to the market in one full-power discharge, or the power can be delivered to the grid with a variable output to provide desirable balancing. The frequency of a storage facility performing this application would obviously depend on the area's load profile but could easily consist of multiple discharges (of a few hours) per day. For this reason, the endurance capacity of the facility is key to determining how often the unit can operate in the market, and larger is usually better.

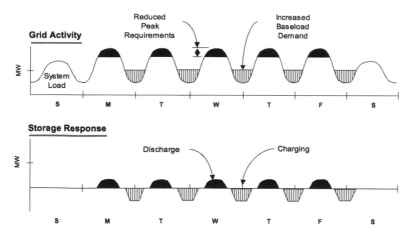

Fig. 4–6. Commodity arbitrage (Ardour Capital Investments).

The key to the profitability (and hence success) of commodity arbitrage obviously rests with the relative difference between peak and off-peak power prices in a region. For example, areas dominated by a single fuel source like the Pacific Northwest (hydro) may not have enough of a price differential in their daily cycle to provide sufficient profitability for a storage facility alone, requiring the unit to be used for other applications as well. Therefore, the relative mix of generation prime movers in the region is a key indicator of success. Regions with particularly high potential for success would have significant unused baseload coal or nuclear generation assets and peak daily prices set by natural gas peaking units. This type of dispatch and price structure would provide significant opportunities to profit on the daily arbitrage cycle.

Three energy storage technologies have the capacity to operate successfully in this market: pumped-hydro storage, CAES, and flow batteries. Only flow batteries do not yet have active installations operating in the market, although the technology has the potential to be scaled to the required size. Market flexibility of these technologies quickly lends them to new roles for the units. For instance, Oglethorpe Power's 848-MW Rocky Mountain (GA) pumped-storage facility was intended to run only in a peaking mode, but in its first calendar year of service (commissioned 1991), the unit ran 364 days in a variety of roles because of its wide operational capability to take advantage of profitable market opportunities. Similarly, Alabama Electric Cooperative's 110-MW Macintosh (Macintosh, AL) CAES facility has also found many uses, operating throughout the day and offering critical support during peak power. All told, the CAES unit can store 2,600 MWh, allowing the unit to operate at full power for up to 23 hours. With the continued volatility and wide spread between on- and off-peak power prices, storage technologies have many opportunities in the evolving power market. However, their siting will remain a central concern, lending support to additional CAES facilities. As opposed to pumped-hydro technology, CAES has much less strenuous siting requirements, allowing these facilities to be placed in relation to the grid's needs rather than its geographical realities. Finally, CAES facilities hold out the potential for large storage reservoirs (tens of hours or more). Such reservoirs could be filled both at night and on the weekend, extending the amount of run-time during the week for the unit.

A final benefit of arbitraging also can be seen from the coal facility's point of view. By acting as a comparable sink for underused power facilities during off-peak periods, these units allow the coal facilities to operate in a more level and less stressful and expensive manner. Because many power facilities are required to ramp up and down on a daily basis to match power-demand levels between on- and off-peak periods, they currently operate inefficiently, increasing their operating costs and environmental emissions. These additional costs can be significant for coal units and especially for gas-fired combined cycle units.

One way to see the potential benefits of additional large-scale storage facilities is to show the impact on a typical coal facility. This example will only deal with annual average costs to show in a broad stroke how large-scale energy storage can assist baseload facilities. It should be remembered, however, that storage can have many other comprehensive benefits improving the coal facility's operations. In this example, we will look into coupling a 1,000-MW coal facility operating at a 71% usage rate with a 405-MW CAES facility (80% system efficiency utilizing a 300-MW compression motor). Operating the compressor for eight hours each night, five nights per week, results in more than 568,000 additional MWh of demand from the coal unit during off-peak periods. This could increase the coal facility's use by more than 6% (to 77%), lowering its production cost by roughly $1.40/MWh (based on $25/MWh, a 2-to-1 fixed/variable split). Therefore, besides the additional $6.7 million of additional power sales the coal unit could capture during off-peak periods, the facility could receive an additional $8.5 million from its current power sales with the new, lower production costs.

Other examples of large-scale storage facilities show that for these units to be truly successful, they must be able to perform a single market role as well as a variety of roles that may change over time. For instance, the 290-MW E.ON CAES plant near Huntorf, Germany, has now been operated successfully since 1978. This plant's market roles have changed according to the situation on the power grid, including as a blackstart unit, a peaking unit (especially in the evening when no more pumped-hydro capacity is available), and as a contingency reserve (because slow-starting coal units are the primary option currently). This ability of the facility to adapt continues, as the CAES unit is now being tasked with compensating for sudden and unexpected wind power shortages as the number of wind turbines increases in northern Germany.

Contingency reserves

To prevent a sudden failure of one or more power facilities or transmission lines from impinging on system stability and security, utilities and ISOs maintain various levels of reserve generation capacity. Different control areas and utilities hold varying amounts of capacity in reserve, as some target a proportion of the system load, whereas others specify a MW level. Overall, these reserves can account for more than 1,000 MW that are reserved and prohibited from generating power. Depending on the response time needed, these reserves are classified as either spinning or nonspinning. Spinning reserves are the first line of response and generally are required to equal the largest unit in the areas. Units providing this service are online, synchronized, and unloaded but ready to serve customer demand immediately should a need occur; all are required to be at full power within 10 minutes. Nonspinning reserves are similar-type units that are held at standby (unsynchronized) or units that are able to start up quickly, with both groups required to be fully online within 10 minutes (nonspinning reserves are simply those not immediately available). Generating units providing these contingency reserves are compensated through capacity payment based in part on the market-clearing price.

As with other ancillary markets, although the technical requirements for contingency reserves have existed since the beginning of the industry, the ISO/RTO competitive market for these services is still evolving. As the definitions of the markets vary across the country, these markets in total can easily represent upward of 5% of the total energy market's value (depending on the ISO). Previously, utilities attempted to harden the power grid to prevent any disruption from occurring. They accomplished this strategy primarily by providing a significant redundancy to essential generating and transmission capacity so that the system would always retain a significant reserve margin. Policy leaders are adjusting that goal slightly to focus on providing a level of survivability in the event (not in case) of a disruptive event. Of course the utilities will continue to strive to thwart an outage in the first place. Significantly more effort is being placed on preventing small, inevitable, localized problems from becoming grid-spanning events, hence the need to find new, better-dispersed resources with greater responsiveness that can be included for these services. The need for these units is especially acute during peak times because units providing frequency regulation services are not permitted to provide contingency reserves at the same time. One increasingly popular alternative solution for supplemental reserves is demand-response programs as they mature from utility-run, dispatchable load–curtailment programs to competitive, opt-in ISO-led price response programs tied to real-time prices and loads.

Energy storage technologies can act as another source of spinning-reserve capacity (fig. 4–7); in fact, hydro and pumped-hydropower facilities are prized because of their superior reaction capability. For instance, if the 1,800-MW Dinorwig (Wales, UK) pumped-hydro facility is held as spinning reserve, the entire plant can reach maximum output in less than 16 seconds. Other existing storage technologies currently provide similar spinning-reserve requirements; for instance the Macintosh

CAES unit can come online within 10 minutes. This unit can also remain operating at only 10 MW (maximum 110 MW), extending its ability to supply a flexible resource. With such responsiveness, additional large-scale CAES storage facilities should also find a receptive market for their capability. In a curious twist of fate, one of the largest proposed energy storage facilities—Ohio's Norton CAES plant (potentially 2,900 MW)—has been proposed for where the 2003 blackout began in Ohio. Had this unit been operating, its supporters suggested the facility could have provided system operators some additional time to respond, helping to prevent or reduce the scale of the resulting event from spreading to such an extent. If other regions follow California to establish reserve adequacy requirements of 15% to 17% planning reserve by 2008, this could provide an additional need for storage facilities to meet these requirements.

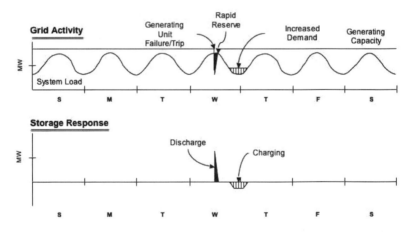

Fig. 4–7. Contingency reserves (Ardour Capital Investments).

Finally, because of the storage facility's ability to shift from charging to discharging quickly, this effectively doubles the unit's system-balancing capability, making it exceedingly useful for reacting to sudden changes in the system load and supply. Pumped-hydro facilities have been used in this way for a long time and remain one of the seldom-realized, but significant, values of these large-scale storage facilities. CAES units can also provide this capability, but compressors and generators are actually separate units, and by operating independently, they provide an additional level of control and responsiveness. Although it may seem counterintuitive to operate the compressors while generating power, this provides the facility with the vital flexibility to provide a number of ancillary services with a lower response time.

Blackstart capability

If the regional contingency reserves are not sufficient to maintain the system and a blackout occurs, the grid operators must undertake the process of restoring power to the grid using units with blackstart capability. These blackstart generating units are able to self-start (or are capable of operating at reduced levels when disconnected from the grid) after the system power is lost; these units are generally combustion turbines and hydroelectric facilities and are the starting point for the system's restoration after a complete system power loss. These specialized plants are needed because once shut down, most generation facilities require system power to begin operation (provide auxiliary plant-load and cranking power for the generator) and reenergize the power grid. Besides self-starting, these units must have the capability to maintain frequency and voltage under varying loads while the system is restored because most of the system inertia will not be online. The entire process can take many hours, so regional transmission organizations (RTOs) such as the PJM require that blackstart generation units have the ability to operate for many hours to restore the system.

The locations of these units are important. To get the main generating units back online as quickly as possible, many are directly tied to the blackstart units or located nearby. Others are dispersed throughout the system at important transmission points for redundancy lest a unit or transmission line failure block the restoration attempt. As utilities have divested their generation assets, the assurance of sufficient blackstart capability in the system is of issue because it is difficult to maintain the capability (necessary equipment, training, and fuel), and there is always the threat that the new owners may not continue providing this service. Without sufficient blackstart resources, system restorations will take much longer, which raises health, safety, and security issues. For these and other reasons, this is not a competitive market but a cost-based service with payment to these suppliers from all customers; there is a benefit to all.

Larger energy-storage facilities (multi-MW, multihour endurance) can assist with system restoration after a blackout or similar event (fig. 4–8). No conventional generating facility is designed exclusively for this role, nor is any storage facility. They would provide this service in addition to their other market roles because there would not be a conflict; also, those storage facilities providing this service would need to have sufficient power and discharge endurance to be of value in the system restoration. Although the location of the units would also be a crucial consideration as to whether they would be suitable as a primary blackstart unit, once the system collapsed, all centrally dispatchable storage facilities would be a resource; those units in less-strategic locations or with lower output capability would simply serve a supporting role for other blackstart units. Therefore, although storage units would in no way displace current power generators providing this role, additional blackstart-capable resources could potentially hasten the system power restoration. Storage facilities already do so in the market (even existing blackstart power generation units rely on battery packs to provide power for their own starting). Besides pumped-hydro facilities, the 290-MW Huntorf, Germany, CAES facility has provided a blackstart capability to some nearby nuclear reactors (besides its other market roles) since the late 1970s. In addition, the recently cancelled Regenesys 15-MW (eight-hour) sodium bromide flow battery was to provide blackstart capability for Innogy's 680-MW Little Barford Power Station in Cambridgeshire, UK, besides providing frequency and voltage regulation for the area.

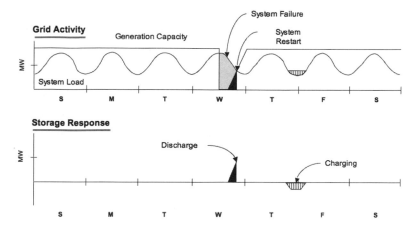

Fig. 4–8 Blackstart (Ardour Capital Investments).

Transmission and Distribution

Energy storage technologies have the opportunity to play a significant role in the transmission and distribution (T&D) market by enhancing the reliability and stability of normal operations and increasing their efficiency by reducing or deferring capital investments. A number of market trends have conspired to make the management of the power grid more difficult in recent years, as well as an area of growing concern for the future. Most recognize that the essential billions of dollars to upgrade the power grid are not simply for additional power lines but for power electronics and other supporting equipment, such as storage technologies, that provide rapid and self-correcting capabilities. Generating capacity expansion is also increasingly oriented toward incremental merchant mid-merit and peaking facilities or distributed resources instead of baseload power facilities, thus requiring far more actionable control capability to be embedded into the power grid to balance the increasing variations and sources of power. With merchant power providing much of the new capacity, however, there is concern that planning the location of these new units is increasingly optimized more for profitable power production (fuel, land, etc.) than grid stability. Storage systems placed at strategic locations could act in concert with flexible AC transmission systems (FACTS) equipment to provide stability and enhance the power grid's operation by providing additional flexibility in operations and planning. In addition, small mobile storage units could offer significant relief until a more permanent solution is developed.

Because their role is to enhance system stability, energy storage facilities designed to support transmission and distribution networks are more concerned with the ability to swiftly inject energy rather than replacing the power generators' role of providing bulk energy to the grid. Consequently some storage facilities could have a relatively small storage capacity while still having a major impact because some roles call for repeated cycling of power in and out of the unit quickly. In the transmission network, large-scale storage facilities (MWs) can provide flexible grid frequency regulation to widespread areas. In the distribution market, however, smaller but faster responding storage units (MWs-kWs) can provide targeted frequency regulation and voltage stability (from their power electronics) as rapidly changing conditions create localized areas of instability. For both markets, the relief provided by the storage at or beyond constrained points in the power grid can reduce or delay required capital investments needed to serve the growing load in a reliable manner.

By acting as both a sink and source of power, energy storage facilities can play a role in the transmission and distribution network along three broad applications:

- **Frequency regulation.** Frequency regulation acts as a stabilizing force to balance supply and demand as the system's load fluctuates. If these changes are not compensated for, damage may occur to both the power facilities and the customers' electrical equipment. Required reaction times span from seconds-level corrections for sudden changes to minutes-long responses for changes in the daily load.

- **Voltage regulation.** Voltage regulation is needed to maintain power flows across the power grid and for the operation of customer power equipment. Reactive power, the force controlling voltage levels, is produced by power facilities and specialized equipment placed throughout the network. Poor voltage control results in additional power loss, especially at the end of a long distribution power line.

- **T&D asset deferral.** Although the infrastructure of the power grid must be constantly upgraded to address the always increasing system load, traditional means of doing so are inefficient, leaving significant capital assets underused. Providing additional resources—both commodity and grid support—in areas with limited transfer capability can help the system ride through peak demand periods, increasing the usage of the network and allowing for a postponement of the required system upgrade.

Frequency regulation

Frequency regulation stabilizes the power grid by managing the moment-to-moment changes in the demand or supply balance of the power grid (fig. 4–9). As load changes, excess generation causes a frequency increase above 60 Hz; insufficient generation causes a decline. Small shifts in frequency (load) do not degrade reliability, but large ones can damage equipment, degrade system efficiency, or even lead to a system collapse.

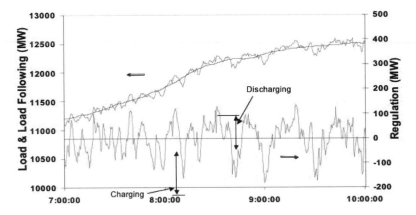

Fig. 4–9. Frequency regulation (Courtesy of Brendan Kirby, ORNL & Consultant).

These changes are first counteracted by the rotational inertia inherent in the connected synchronous generators. As the variation continues, regulation can also be provided through generating units operating under automatic generator control or participating in manual frequency control, both of which can change output quickly (on the order of MWs per minute). Automatic generator governors restore frequency if it deviates more than 0.036 Hz from normal. If the variance in frequency continues to grow, the power facility may lose its synchronization from the grid at ±1.5 Hz to prevent significant equipment damage, which can begin to occur at ±3.0 Hz. Besides the obvious costs of damaged equipment, variations in power output can also produce raised operating costs in plants constantly altering output, plus the increased operation and maintenance (O&M) costs from the added wear on the unit, which also reduces its operating life. Frequency regulation is often a greater issue on small or isolated utility systems, because system inertia on a larger power grid provides more time for generator governors to respond. Having less system inertia, the frequency will fall faster on a small grid system in the event of power supply imbalances. Therefore, these isolated power grids must be capable of a far more rapid response to frequency deviations, or the frequency change will begin to affect equipment throughout the system more rapidly than on a larger, integrated system. For some smaller utility systems, regulation is combined with spinning reserves to make the most of scant excess capacity.

Load following is similar to regulation but on a longer time horizon. Rather than focusing on the minute-to-minute load variations required of regulation, load following provides energy correction on an hourly basis for any mismatch between energy supply and the system load. Although useful throughout the day, this service is obviously more valuable during the morning pickup and evening drop-off, when the greatest changes occur. As the load for the area changes, the system operator must bring on more generation to meet that rising demand, which can change 10% per hour. These power facilities are forced to ramp up generation according to the demand level.

Frequency regulation services, as well as load following, are currently produced by power facilities. For example, PJM currently designates 1.1% of the projected peak capacity to provide regulation and load following for the system. RTOs like the PJM have developed the service into a competitive market so anyone could bid in to provide the service; however, frequency regulation is a technical requirement of the system, so all utilities need a means to provide it. In other areas not under an ISO/RTO structure, the utility provides this capability to itself, but as it is rolled into operations, breaking out the cost can be difficult—even for the utility.

Some energy storage facilities are well suited to provide frequency regulation; in fact some large-scale storage facilities are already key providers of these services. Pumped-hydro and CAES facilities have provided these capabilities for years (decades) and are sometimes even the primary providers of regulation for the entire utility. These facilities counteract the frequency variations by increasing their discharge to meet a negative imbalance in the system load and decreasing their discharge to meet a positive imbalance in the system load, similar to how power generation facilities currently provide the service. Many of these storage units can also provide regulation services during their charging episodes by reversing the previous actions. Because of the large scale of these facilities, these units can also provide longer charges and discharges for load-following duties.

Other storage technologies are able to provide frequency regulation in multifunctional facilities. An early example was the Puerto Rico Electric Power Authority's 20-MW, 14-MWh lead-acid battery storage plant at its Sabana Llana substation, which was installed in 1994. This $21-million facility provided spinning reserves, frequency regulation, and voltage control services for the island. These capabilities proved extremely useful during the fall of 1998 in the aftermath of Hurricane George, when unreliable power supplies and transmission reliability caused a number of load-shedding events and increased spinning reserve requirements— requirements that the battery facility reduced during this time of scant resources. A more recent example is the recently inaugurated 46-MW NiCd battery facility for the Golden Valley Electric Association (Fairbanks, AK). This unit supports the radial makeup of the Golden Valley's power grid, providing frequency regulation and lowering the larger generator spinning reserve requirements common among Alaskan utilities because of their nonconnected status. These energy storage units provide frequency regulation by repetitively cycling from charging to discharging in order to balance a mismatched system load. This is a different tactic than existing generation facilities that provide this balancing role through a constant change in output level. Because these smaller units are not sized to the storage capacity of these other units, these smaller storage facilities must have fast reaction capability and a high cycling limit to effect a similar result. To maximize the effectiveness of the storage facility, the facility should also have as deep a discharge capability as possible.

Although these multifunctional storage facilities have been useful (especially in small power grids), providing frequency regulation requires a high cycling capability of the storage facility, limiting the number of storage technologies actually available to provide this service. By nature frequency regulation is comprised of essentially a constant need to charge or discharge the storage unit, and across the 20-year life of a storage facility, providing the service can easily require more than 500,000 full charge/discharge cycles (based on a 15-minute cycle) if constantly engaged. Power generation facilities of many types currently provide this service, but their performance is much slower than optimal because their response time is not capable of reacting to all of the constant changes; and attempting to match what is actually needed imposes a life penalty from additional wear (in fact, the current level of operation does as well). Therefore, these units' control schedules are purposefully slowed to reduce the life impact at the expense of performance. One solution to provide a much-needed *prompt-response* capability for regulation service is the proposed 1-MW, 250-kWh Beacon Power Smart Energy Matrix (SEM) grid stability facility, which is designed to operate purely in the regulation role. Matt Lazarewicz, vice president and chief technology officer of Beacon Power, notes in a personal correspondence, "Beacon Power's SEM is particularly well suited for frequency regulation, and is much more effective than generators used today. The SEM is purposely designed for fast response (milliseconds versus minutes for generators) with high reliability, much lower operating costs, and operates without any emissions." The flywheels in the unit possess many advantages for this type of operation; they have the required life cycle, deep discharge, and fast response rate needed for this role. Because the system is designed only for this high-cycle role, its performance and operating cost can be much better than that of a generator for this role. Finally, because this unit is enclosed completely in a typical container shell, the unit is mobile and able to be reassigned to different areas of the grid as warranted when conditions change.

Voltage regulation

Maintaining and regulating the system voltage level is important for efficient transmission of power and customer requirements. Low voltage conditions arise on power systems from two main sources: highly loaded power lines and long, unsupported distribution lines. Uncompensated for, both of these situations can negatively impact the amount of power able to be transmitted through the power line. As with frequency regulation instabilities, swings in customer power use impact system voltage, particularly with the use of heavy motive equipment. Unfortunately customer requirements for voltage stability are increasing, in part because of the increasing penetration of electrical equipment into the commercial and industrial setting. According to EPRI, most distribution-system problems are voltage sags of 10% to 30% below nominal, extending from 3 to 30 cycles in duration.[1] These types of disturbances are responsible for causing motor-controlled manufacturing lines to shut down or trip off-line. Although some consumer electrical equipment can operate within a 10% range of the rated voltage, much electronic control equipment or information technology produced now has a significantly lower limit (5% or less) before the equipment is affected or damaged. With continual distribution-system expansion, maintaining sufficient voltage control will remain a challenge; ongoing budget constraints limit the ability of the utility to install all the needed voltage control equipment.

To control voltage levels, reactive power is produced and managed throughout the system. Reactive power, measured in mega volt ampere reactive (MVAR), helps control voltage levels and is required to power consumer electrical equipment as electric motors and the like for creating magnetic fields. Reactive power primarily is produced at power facilities through voltage regulators, but it is also controlled with specialized equipment (capacitors/inductors, Static VAR Compensators [SVC], transformers, etc.) spaced throughout the network that adjusts the voltage level as needed. Spacing the equipment throughout the network is one way the industry currently manages the reactive power levels to prevent small incidents from growing into larger, area-wide events. Unfortunately, by producing reactive power, the generating facility lowers its ability to produce real power for sales—crucial during peaks when reactive power demands rise along with the need for power.

The reactive power needs of the system operate dynamically with the use of the transmission system; if they are not addressed, the imbalance in reactive power levels will negatively affect the carrying capacity of the power line. When a transmission line has only a low load, the capacitive effect dominates, requiring absorption of reactive power. Conversely, at high loading the inductive effect dominates, requiring the production of reactive power. This relationship is nonlinear, so periods when the power line is fully loaded require significantly more reactive power injections to prevent a voltage loss than when it is only at an average load. There are also problematic, significant drops in voltage at the end of long distribution lines, especially during high-load periods. Both of these situations require the utility to inject reactive power during high-load periods. With improved carrying capacity at these congested points, the entire carrying capacity of the power grid can be improved.

Fast-reacting energy-storage technologies are well suited to counteract the variations and instabilities in the voltage level that occur every day in the power grid. This is accomplished by providing direct reactive power from their associated power electronics and real power (because of the dynamic relationship between reactive and real power). Energy storage facilities designed to provide support for the grid are concerned with quickly injecting stability into the grid. They are therefore designed with a smaller endurance capacity but contain a faster reaction time and deliverability capability instead of long-term endurance (fig. 4–10). Similar to other voltage control equipment, these facilities are not envisioned as replacements of power facilities (the main source of reactive power) but as distributed providers of reactive power to correct a momentary imbalance in localized areas of voltage instability. This role is especially critical during periods of peak power demand when voltage instability will be worse in the remote areas of the power grid.

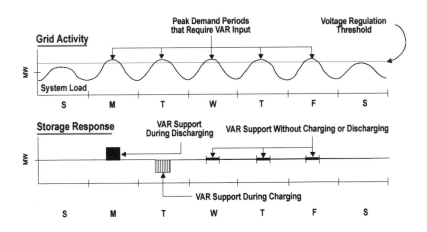

Fig. 4–10. Voltage regulation (Ardour Capital Investments).

Early examples of energy storage technologies providing voltage stability were limited to a handful of battery-based, multifunctional systems such as Puerto Rico's Sabana Llana substation facility, and PG&E's substation facility in Chino, California. More recently other storage technologies with faster reaction times and a far higher cycle life have become more common and are expected to become more widespread as utilities often simply want a more direct alternative to voltage control equipment. Another aspect of these alternative energy storage solutions is that they are trailer-mounted mobile units. This can be an important attribute to the unit; areas of voltage instability normally grow in areas previously not suffering from these problems because of load growth and usage change by consumers. With its significant experience with energy storage technology, Detroit Edison installed a demonstration zinc-bromide-based, trailer-mounted 200-kW, 400-kWh flow battery from ZBB Energy Corp. The unit was able to provide relief for a number of load management roles, including frequent voltage sags. Because the system was truck mounted, it was extremely mobile, allowing for the quick redeployment of the system to other locations.

Raising the usage and reliability of an existing transmission network through improved voltage control equipment is frequently a more cost- and operational-effective option for expanding the carrying capacity of the transmission line. Besides such experience at the Wisconsin Public Service installation of the American Superconductor D-SMES (Distributed Superconducting Magnetic Energy Storage) units, Entergy has continued to operate a number of similar D-SMES units since 2002 in different areas throughout the Houston, Texas area. These units were installed in substations to solve voltage-related problems and to increase the power-transfer capability for critical transfer paths in congested areas, improving the carrying capacity of the entire area. The need for these units is driven by seasonal peaking loads from increasing summer cooling requirements, and it was a cheaper solution than siting additional generating capacity—the other option considered.

Transmission and distribution asset deferral

Providing utilities a means to better control the timing of their distribution-system upgrades would be a boon to firms struggling to provide increasing reliability with diminishing resources. One strategy is to selectively defer a portion of these required upgrades—especially in the lower-voltage transmission and distribution where existing problems from underinvestment increase in incidence. By providing an additional five or more years of usability to the existing transmission equipment, scant resources would be available for other purposes; but more importantly, it would allow the system upgrades to be planned around utility's schedule, not the other way around. The need for finding alternative solutions to this problem is apparent because the drives for them—continued demand growth and aging equipment—show no likelihood of abatement; in fact, these point to even greater demand for alternatives. Based on data from 1999, upwards of $12.7 billion are added to the capital infrastructure base each year: $2.3 billion in the transmission market and $10.4 billion for the distribution market.[2] This is needed to keep up with demand; over the last 10 years, demand growth in the United States has increased 20%, including a shift toward higher reliability. Meanwhile, the existing equipment continues to age, with much of the existing equipment approaching or surpassing its design life.

With this growing demand, distribution lines and their associated substation equipment require continued upgrades. Here again the reason is for reliability, but long-run economics dictate how the upgrades are completed. Because upgrading power lines entails expensive construction costs, the conventional approach for utilities is to upgrade only when a particular power line nears its transmission capacity more than a few days per month. Unfortunately, demands on these lines are often extremely variable, leaving such a power line with an overall low usage. At this time, the utility will raise the carrying capacity of the line by about one-third, a level sufficient to preclude another upgrade for a while. Operating in this way without a storage component, the power transmission and distribution system must obviously be built out, making it capable of meeting the greatest demand under harsh conditions. It is evident, however, that it is not the entire system that reaches constraint but single points in the system.

Storage technologies can help alleviate both the lack of transmission capacity and the stability problems that ensue during periods of high demand (fig. 4–11). Sited past a bottleneck, they can provide a prepositioned source of energy, which will allow the system to ride-through a few short-term peak demands during the month—postponing the need for an expensive upgrade on the line until a more sustained level of demand warrants an upgrade. As an added benefit, the power lines will experience greater, but less intense, use and wear through reducing overloaded lines (peak power transfers that rise above rated limits incur far more wear and tear on a line than pushing additional power through the line at night when the usage is low). A key strength storage facilities have over traditional solutions to counter reliability and capacity problems is the multifunctional capability of the unit. Frequently, it is not simply real power that is needed, but also voltage control and even frequency regulation services that are needed. By providing such system stability support, storage technologies can also help avoid the need for more expensive power stability and control equipment. Evaluation of storage technologies for this purpose has been undertaken for years, with countless surprising findings in storage's favor. PG&E even concluded back in 1994 that a 1-MW, 2-hour duration battery storage system (with a 10-year operating life) priced at $700 per kW could enable the profitable deferral of one additional substation each year.[3]

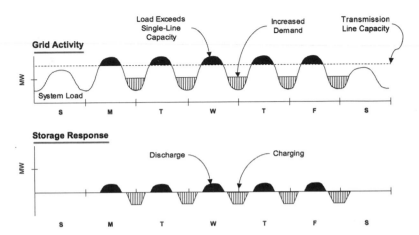

Fig. 4–11. Transmission and distribution (T&D) asset deferral (Ardour Capital Investments).

At the distribution level, other utilities such as PacifiCorp are beginning to use other storage technologies to postpone remote power line upgrades—a constant challenge that is both costly and lengthy. Long-distance, low-voltage distribution lines are especially prone to low-voltage problems. Here, frequent upgrades are either impractical or impossible, giving greater support for an energy storage technology solution. In this case, PacifiCorp has a 200-mile, 25-kV distribution feeder near Moab in Utah, where increasing episodes of high loading caused periods of severe voltage problems. The problem had become so severe that new connects were delayed, and power quality problems were so numerous that many consumer complaints were lodged with the Public Service Commission. Several potential solutions were examined, including conventional planning options such as line reinforcement, additional power lines, substation upgrades, and increased reactive compensation. However all were found either impractical (due to environmental restrictions) or expensive. The use of energy storage as an alternative proved to be the most economically attractive solution and resulted in the installation of a 250-kW, 2,000-kWh vanadium redox battery energy storage system (VRB-ESS) with a vanadium-redox flow battery from VRB Power Systems in late 2003 on a distribution feeder near Moab, Utah.

Continuous full-power, daily cycling operations began in March 2004. Feeder voltage deviations have improved by 2%, and feeder power factor improvement from VRB-ESS have reduced line losses by 40 kW, more than offsetting the parasitic losses of the VRB-ESS. The unit is capable of providing a variety of functions, including peak shaving, load following, dynamic frequency control, voltage support, and islanded operations; and it provides premium power for industrial and commercial customers on the power line. Although the final solution has always included developing a power line upgrade, by using the energy storage facility, the utility now has the ability to schedule the work at its convenience, while power quality along the line is assured. According to Brad Williams, director of Power Delivery Business Technology at PacifiCorp,

> The VRB application buys time to work with communities and regulators on difficult siting and aesthetic issues that new sub-transmission lines and substations always seem to bring. Once the permanent sub-transmission line and substation solution is built in a number of years, we can relocate the VRB to another site for additional benefit.

Retail Market

Although energy storage technologies are already widely used in the retail market, they stand to play a much wider role protecting consumers from growing power-quality issues or volatile power prices. In this way, as opposed to the revenue generation opportunities of the wholesale power market, the retail market for energy storage technologies is more focused on providing cost-saving and loss-prevention capabilities. Commercial and industrial firms have enacted programs to reduce their cost of using electricity, both the direct cost of the service and the impact it can have on their business operations, for many years. U.S. firms spent $133 billion for electricity service in 2000.[4] Unfortunately, those firms with larger and more unpredictable demands are normally subject to a demand charge and multistage tariffs, increasing the propensity for a greater variability in their costs; much as these firms dislike large power bills, however, they like unpredictable bills even less. Conventional strategies to reduce these firms' exposure have picked much of the low-hanging fruit, but some strategies threaten to interrupt production schedules, costing more to the firm than what can be saved from lower energy bills. Protecting ongoing operations from poor power quality is also a growing issue for many firms, as increasingly sophisticated IT equipment and computer-controlled process machinery grow less tolerant of voltage or frequency variations. In fact, there are some estimates suggesting that poor power quality now costs the U.S. economy $150 billion annually from damaged equipment or lost production and work time.

Just as changing market conditions drive the need for more storage technologies, improving storage technologies expand the number of applications able to be addressed. Many of these technologies will be increasingly implemented into firms' operations where existing storage solutions do not currently exist, but the need is growing. For shorter-term power-quality problems, fast-reacting energy storage assets like UPS units can improve the quality and usefulness of the customer's power to protect critical loads from any disruption that would impact their normal operations, or even provide short-term ride-through supplies to bridge power supplies until on-site power supplies can start up and protect critical loads. To support cost-minimizing strategies, these technologies can also provide multihour discharges to reduce demand charges and manage energy usage. Growing interest is also emerging to take advantage of the cycling capability of storage technologies so as to balance the load from short, repetitive motion.

By acting as both a sink and source of power, energy storage facilities can play a role in the retail market along three broad applications:

- **Power quality:** Provides a means to protect sensitive information technology or high-speed industrial equipment. On-site UPS equipment can also act as a bridging power source until backup generators come online in the event of an interruption of system power.

- **Energy management:** Provides an on-site cache of energy to modify demand profile. This cache can either support a cost-of-service minimizing strategy through some form of peak shaving, or provide a regenerative energy charge/discharge capability for heavy cyclic loads to reduce wasted energy.

Power quality

As mentioned previously, poor power quality is a growing problem for U.S. businesses. Although the exact value is difficult to determine, many current estimations of the annual cost to the U.S. economy from poor power quality agree with the DOE's value of $150 billion from interruption or loss in operations, work time, or damage to increasingly expensive equipment. This problem has been acknowledged for many years, prompting many utilities to estimate the impact of this poor power quality on the economy in the areas they serve. Duke Power (now Duke Energy) developed a detailed estimate of these customer impacts in 1996 and found that voltage sags and interruptions cost $4.3 billion per year to their customers. The DOE extrapolated this value to the national level, and derived a value in the range of $150 billion in total costs.[5] Unfortunately, these estimates will always remain subjective, with some estimates pointing to an even larger problem. For instance, one early survey of 450 information systems executives at Fortune 1000 companies in the late 1990s reportedly revealed that power quality problems resulted in lost data and productivity that were estimated to cost U.S. businesses $400 billion a year.

Both industrial and commercial firms suffer from poor power quality conditions; while transient power fluctuations disrupt and damage industrial batch processing (lost work product) and continuous processing (schedule disruption), commercial firms are also feeling the impact as fault-intolerant information technology spreads throughout the firm. Voltage sags and swells are the most common power quality problem, and these events can originate from obvious external sources, such as lightning strikes, but also from less-obvious internal sources, such as large electrical loads (motors, etc.) suddenly switching on and off. Surprisingly, most power disruption events do not last long. An EPRI study analyzing utility disturbances found that 98% of all events last less than 30 seconds, and more than 90% of power quality disturbances last for less than 2 seconds.[6]

In the industrial market, the effects of poor power quality are well known. The same EPRI study on chronic U.S. retail power quality issues revealed that more than 90% of all manufacturing facilities would experience utility voltage sags of greater than 20%, with more than 30 dips per facility (greater than 10% each year). Serious voltage sags lasting only a few cycles with a deviation of 30% from normal have been known to trip motor drives, controllers, and other industrial equipment, causing significant interference with the manufacturing line. Although most discussions about power quality have dealt only with power from the utility, power quality within a large manufacturing facility can also vary and be affected by the alternating uses of electrical equipment throughout the plant. Motors and drives for precision-controlled equipment are increasingly susceptible to voltage disturbances, even from those stemming from within the facility itself.

Commercial firms are also requiring an increased level of power quality protection. First, as mentioned previously, information technology in the commercial market is also increasingly susceptible to poor power quality; other findings from EPRI show that a typical computer system annually experiences around 300 power disturbances outside the manufacturer's voltage tolerance limits, potentially disrupting millions of dollars of transactions for some financial institutions. Secondly, the type and location of the commercial building can have an impact on the power quality available. For instance, buildings in older parts of a city may experience a higher failure rate because of congestion and aging equipment. Unfortunately, the suburbs do not offer a much better solution because many commercial office parks are widely distributed, with older ones built to carry a far lower load than is currently demanded.

With a greater distance from the local substation (compared to an industrial facility), older equipment on a longer, smaller circuit poses a growing need for stabilization solutions.

What these firms want—either from their service provider, or increasingly from their own energy management strategy—is a means to provide a capability for loss-prevention from these transient frequency variations or voltage surges and sags. UPS systems, a long-standing use of energy storage technologies, have provided both enhanced power quality and reliability to consumers for many years (fig. 4–12). Its first use is to protect critical loads from momentary power sags that damage sensitive equipment. As the impact of poor power quality grows, the global market for UPS systems was estimated to be $7 billion in 2002, with growth over the foreseeable future to average 7% a year.[7] Because most line instability incidents that disrupt production lines or computer equipment last only a brief instant, storage facilities can provide *clean* power by providing repeated, shallow discharges to combat such issues as voltage dips and sags. Often these systems are made up of lead-acid batteries, but other storage technologies are extending the capabilities of this class of equipment. A second important use of a UPS system is to provide a ride-through or bridging-power source of power in the event of an interruption of utility service to critical loads until an on-site power generation is running and stable. Generally, 15 to 20 seconds are needed to either bring a backup power generator online or a switch to a different feeder line.

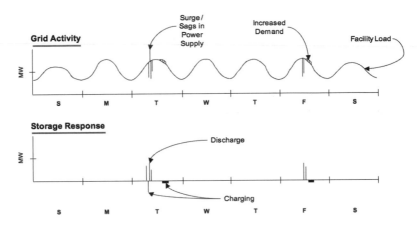

Fig. 4–12. Power quality (Ardour Capital Investments).

Finally, once a backup generator begins supplying power in an off-grid environment, power quality is still a concern. Simply put, firms want similar power quality from their distributed generation setup as they get from utility service. Unfortunately, most do not. Fuel cells, some natural gas-fired units, and other smaller-generation technologies are unable to reliably react quickly to a changing load. If the generators serve only a nonvarying load for the facility, the generator may handle most variations in load. If the load covered by the backup generator changes because of processing requirements, however, some storage technologies such as flywheels show promise to provide a stable (frequency regulation and voltage stability) environment while the facility is isolated; in essence, the storage unit would provide a load-following capability because the single generator has a far more difficult time responding to a changing load than the utility system power. To perform this role, the storage asset will need the ability to operate in dynamic conditions with many charge and discharge cycles during this period of interrupted utility service. Because the storage unit can react more quickly, these combined systems will be able to perform over a much wider range of operating conditions than just the distributed generation asset itself.

An outsized example of the need and usefulness of a UPS system can be seen in the power quality problems solved at a semiconductor wafer fabrication facility in Phoenix, Arizona. Utility power disturbances are a great concern in the wafer fabrication industry because producing the final product can take 30 to 50 days, and disruptions at any point along the process can ruin the entire batch. To add an additional level of protection, in August 2000 STMicroelectronics installed a 12,500-kVA PureWave UPS system (five 2,500-kVA units) from S&C Electric Company operating at 12,470 volts (V). Based on the analysis of Arizona Public Service's system events, three options were considered: numerous conventional 480-V UPS systems installed within the facility to protect critical process; a medium-voltage, high-speed source-transfer system installed between the plant's two incoming utility feeders; or an off-line PureWave UPS system installed at the utility source. The single medium-voltage UPS system was ultimately selected because it provides protection for the entire operation and ensures a higher reliability because a fault on an unprotected aspect of the facility could affect the entire process. This off-line system operates only during a utility system disturbance and switches to the battery bank within 4 milliseconds (less than 1/4 cycle). When the system power returns to normal, the load is then switched back. Because the duty cycle is only expected to be 100 seconds of run time, such a high power rating can be obtained from a relatively small system. The system has proved its worth since its installation; it has protected the facility more than 90 times in the first four years of operation—the first a mere 10 days after installation.

Energy management

Managing energy usage at commercial and industrial firms has become far more involved than simply signing up for an energy audit. As mentioned previously, the original desire by commercial and industrial firms to simply have lower energy bills has been joined by the aspiration to prevent volatile power prices from affecting the bottom line. For many firms, even a small rise in energy costs could wipe out hard-won efficiency improvements; worse yet is simply an unpredictable price. Conservation programs have been used for many years to reduce enterprise energy costs, but they have two drawbacks: many do little to counteract the specific high-use periods that drive demand charges, and those load-shifting programs that do quickly begin to interfere with production operations. These demand charges are commonly a much greater factor in many industrial heavy users of electricity, especially if their energy use fluctuates from batch job processing.

Like conservation programs, energy-management programs using storage technologies would target reducing power-intensive activities that occur during peak periods; unlike most conservation programs, there could be little interruption in the operation of the facility. These technologies can help commercial and industrial firms reduce their costs along two main routes: peak shaving and providing a sink to capture otherwise wasted energy in repetitive motion activities. For both uses, storage augments and expands the capability of any existing energy management program. Because customers with more stable energy use patterns are generally more welcome to a utility than those with highly variable loads, the wider introduction of these technologies into the marketplace could even gain support from the local utility.

Peak shaving. Peak shaving allows customers to directly reduce their utility bills by arbitraging power stored from off-peak to on-peak periods. By doing so, energy storage technologies allow firms to engage in load shifting without the subsequent effects on the production schedule. The total use of energy would not be affected, but the firm's costs would decrease because of the lower use during higher-priced periods of the day (if the firm's tariff has multiple levels). However, the main focus for such actions is generally to reduce the firm's demand charge (fig. 4–13). By maintaining such a cache of energy on-site, the energy storage facility can also be integrated with other energy management strategies gaining favor, such as a demand response program, to provide the firm with greater flexibility to reduce its usage cost of energy. To date, thermal energy storage technologies already have a well-established track record of successfully reducing demand charges at industrial and commercial firms, providing an example for other technologies to follow.

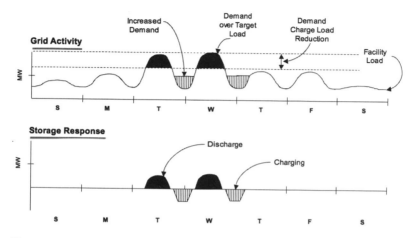

Fig. 4–13. Peak shaving (Ardour Capital Investments).

Although heavy energy use increases the commodity energy charge, highly cyclical loads such as cooling or batch processing can increase the demand-charge component of the bill dramatically. For this reason, even for many firms where overall energy costs are only a small part of general costs, often certain components of their energy usage may be well-suited to incorporating an energy storage component. For example, the DOE notes that 25% of commercial usage for electricity is for cooling load, a highly cyclical operation. One current example where storage technologies are already used for peak shaving roles is thermal energy storage (TES), which assists commercial chillers to reduce the cooling load of the building. These central chillers predominate in medium- to large-scale buildings (the larger the building, the more potential benefit from TES), and 80% of buildings greater than 200,000 square feet use them, about 25% of all buildings.[8]

For customers with existing air conditioning equipment, thermal energy storage systems can be justified as a retrofit based on the reduced costs of using off-peak instead of on-peak power. Typical payback periods for these retrofit thermal energy storage systems can range from one to three years. This is possible through savings by reducing the demand charge and lower peak-power purchases. Following this strategy, thermal energy storage facilities can reduce peak power demand by 50% and reliably reduce cooling load costs anywhere from 20% to 30%. Because each retrofit installation is project specific, exact costs are difficult to establish, but generally range from $250 per peak kW shifted to $500 per peak kW shifted.[9]

Integrating a thermal energy storage unit with new medium to large construction provides for even greater benefits. In most cases the decreased design requirements on the building's cooling system will pay for the cost of the system, plus provide the additional operating cost reduction described in the retrofit installation. In particular, because TES systems allow a building air conditioning system to operate at lower temperatures, duct sizes can be reduced by 20% to 40%. Because the cooler air also requires a smaller volume of air to cool the building, fan motors, air handlers, and chilled water pumps are smaller and less expensive. Based on the reductions in capital equipment needed when a TES system is integrated into the original design, EPRI has estimated that overall ventilation and air conditioning costs are reduced by 20% to 60%.[10]

Besides these costs, energy storage technologies can provide additional support and savings. Hotels, hospitals, and other commercial buildings with large cooling loads require significant power demands during start-up—events that can occur many times per day. In these situations, power constraints often require staggering the start-up of cooling equipment; an inconvenience that costs time and reduces the flexibility of operations. With additional energy storage technologies to provide much-needed short-term pulse-power to reduce these intermittent demand spikes, anecdotal evidence suggests that peak demands could be cut by upwards of 25%.

Regenerative energy. Another area of retail energy management gaining significant attention of late has been the use of an energy storage technology to improve the efficiency of repetitive motion operations. These operations are often extremely wasteful; for some operations, strategies have been devised, such as the counterbalance in elevators, to recoup otherwise wasted energy, but many operations remain without satisfactory solutions. Energy storage technologies with high cycling capabilities hold out the potential to play a much wider role in capturing and reusing otherwise wasted energy in repetitive motion processes (fig. 4–14). This role of acting as a dynamic sink and source for power fits well with transportation sector applications, where repetitive starts and stops produce inefficient use of energy. Although container port lifting cranes and light-rail/subway systems all exhibit these usage patterns, many other applications could benefit from such an addition to the process, even including the manufacturing floor, where voltage fluctuations can be caused by the many motors turning on and off constantly throughout the day. It should also be noted that if an energy storage solution is used to lower the operator's costs for the customer, even the local utility stands to benefit (especially on larger installations) because the load swings are reduced on the power lines feeding these installations, especially in the case for urban subway/electric bus systems where the extremely variable (and short-duration) demand requirements of the starting and stopping often place unacceptable strains on the city's power system and actively work against the goal of providing reliable, stable power to other customers. Although battery technologies generally dominate the retail energy market, traditional solutions such as lead acid batteries are not able to support these applications because of the high cycling requirements. For many of these more demanding roles, flywheel systems have seen some recent success.

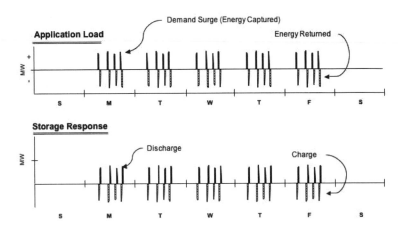

Fig. 4–14. Regenerative energy (Ardour Capital Investments).

For instance, urban rail systems represent a key opportunity for storage technologies to capture and recycle energy usually lost during braking. A prominent example is England's Urenco Power Technologies (UPT) deploying a 400-kW flywheel system on the San Francisco MUNI rail system, enabling upward of a 20% reduction in energy purchases through capturing the braking energy of the trains instead of allowing that energy to be dissipated through braking. Besides the energy saved, the stored energy will reduce the energy demand spike required during the acceleration phase, creating less wear on the equipment. Besides San Francisco, New York City's transit system has also investigated the use of flywheels from UPT. Here, a 1-MW UPT Trackside Energy Management System has operated since January 2002 at the New York City Transit (NYCT) Far Rockaway Test Track, providing voltage support and optimizing the use of a new generation of trains by reduction of peak power demand through recovery of the braking energy. A solution was needed because voltage sags of 10% or more were common during acceleration of these new trains. The candidate technologies included batteries, super capacitors, and flywheels. Batteries were rejected because they could not meet the arduous cycling requirements of the application. Super capacitors were also considered unsuitable for the application as the cyclical demand greatly reduced their life to 5 years or less under these conditions. Flywheels were selected because of their high cycling capability, long life (20 years), and minimal maintenance. The flywheels were able to stabilize the voltage of the system and also supported the adjacent operational lines during normal operation.

In the previous example, lowering demand spikes helps reduce power fluctuations on a utility's highly congested urban power grid. Using energy storage technologies has benefits for rural utilities as well (besides the industrial firm). For example, in the Usibelli Coal Mine in Alaska and its utility, the Golden Valley Electrical Association (GVEA), the mine installed a $1-million, 40-ton flywheel (in agreement with GVEA) to support the mine's dragline load. During the unit's operation, the mine's electrical load is greatly affected by the dragline because its demand alone can swing by as much as 4 MW (across an 8-MW band) and cycle every 60 seconds. To provide a more manageable power level to the utility (and hence, lower their costs), the mine installed a flywheel system that can produce 5.2 MW for 3 seconds (plus a 1.8-MW generator for longer-term energy needs), thereby reducing the worst of the demand spike. This set-up was necessary because GVEA's turbines could not handle the large power swings of the dragline, causing the power frequency to fluctuate with each dragline cycle. With the use of this flywheel system, the average power demand from the utility is now only around 2 MW to 2.5 MW, and GVEA only sees about a 500 kW band of power fluctuation. As a precaution, there is a radio link between the flywheel building and the dragline, so the dragline operators are warned when the flywheel shuts down for any reason. Without the flywheel in operation, each use of the dragline can be watched on the utility's meters in Fairbanks (100 miles away), and even noticed in Anchorage (350 miles away).

References

1.   *EPRI distribution system power quality monitoring project* (DPQ study). 2003. Palo Alto, CA: Electric Power Research Institute; Knoxville, TN: Electrotek Concepts.

2.   Edison Electric Institute. 2001. *Getting electricity where it is needed.* Washington, DC: Edison Electric Institute.

3.   Abbas, A., S. Swaminathan, and R. Sen. 1997. *Cost analysis for energy storage systems for electric utility applications* (SAND97-0443), 14. Albuquerque, NM: Sandia National Laboratories.

4.   Energy Information Agency. 2004. *Annual energy review 2002.* Washington, DC: U.S. Department of Energy.

5.   Swaminathan, S., and R. Sen. 1998. *Review of power quality applications of energy storage systems* (SAND98-1513). Albuquerque, NM: Sandia National Laboratories.

6.   *EPRI distribution system power quality monitoring project* (DPQ study).

7.   *World UPS markets: Alternative energy storage solutions.* 2003. San Jose, CA: Frost & Sullivan.

8.   *Thermal energy storage—Economics and benefits.* 2002. Arlington, VA: E3 Energy Services, LLC.

9.   Ibid.

10. Electric Power Research Institute. July 1991. *Cold air distribution with ice storage.* brochure CU-2038. Palo Alto, CA, EPRI.

5 RENEWABLE ENERGY AND STORAGE

Increasing the use of domestic renewable energy resources is one of the primary goals of U.S. energy policy makers. The U.S. federal government has led this effort with its continued support of the National Renewable Energy Laboratory's National Wind Technology Center research and the 1.8¢ per kWh Production Tax Credit (PTC). Unfortunately, the sustainability of this support has been less than reliable—consistency of support is crucial for market development, and it has been lacking at this level. Luckily, state governments are joining the fray through their Renewable Portfolio Standards (RPSs), which are increasingly seen as the primary driver for the deployment of additional wind turbines. Increasingly, these RPSs are setting more aggressive targets, as in California, where 20% (possibly moving to 33%) of all generation is required to come from renewable sources of energy by 2017. Building off this newfound support, there are increasing calls for renewable energy (in Europe and the United States) to provide 10% of all electrical generation (or all energy usage) by 2010; 20% by 2020; and 50% by 2050.

Unfortunately, there is an underlying problem to this growth target. Renewable energy market penetration continues to lag expectations, and significant efforts on many fronts will be needed to expand the level of renewable energy generation to the point where it can begin to account for such a significant segment of the supply base. According to the U.S. Department of Energy (DOE), renewable energy sources (nonhydroelectric) provided only slightly more than 2% of the nation's electric power in 2002, holding even or slipping each year since 1991, when it peaked at roughly 2.5%. Including hydroelectric power simply moves these percentages upward to 9% and 11%, respectively, but the trend continues to decline. Shifting to a total energy accounting makes little difference, as renewable energy resources account for only 6% of all energy consumed in the United States in 2002, down again from more than 7% in 1991.[1] These problems extend into the future. As can be seen in figure 5–1, there is a growing deficit in the level of expected renewable energy production and the previously stated goals for such renewable energy. Much of this shortfall stems from lack of additional resources from traditional renewable energy sources, such as hydroelectric, biomass, landfill gas, and the like. Although power production from these resources will continue to expand, only solar and especially wind power are flexible enough and can be integrated most easily into the existing infrastructure to hold out the prospect of meeting this or any other significant renewable energy production increase.

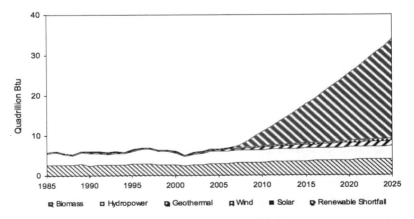

Fig. 5–1. Renewable energy shortfall (Data: DOE, Annual Energy Outlook–2004).

Wind power has long been sought after as a source of renewable energy, but its cost and operational issues have retarded significant growth in the United States until recently. Fortunately, the wind energy potential in the United States is immense, with proponents of wind energy quick to point out that the continental United States has the potential to produce nearly 11,000 billion kWh of electricity—three times current demand of the country.[2] For wind energy project developers, however, even more interesting is that areas with the highest sustainable winds are highly localized, with only a small percentage of the contiguous United States containing the majority of the addressable energy resource. Developing these resources is accelerating, and more than 6,000 MW of wind turbines were installed in the United States by the end of 2003—half of this total within the three preceding years. These advances have only become possible as wind energy has grown to be far more economically competitive—essentially matching (or beating) the production cost from competing power sources if located in favorable wind conditions. However, even as production costs continue to decline, inherent operational disadvantages will continue to put these wind power turbines at a significant disadvantage in the broader power market.

Coupling energy storage technologies with wind turbines can solve many of wind power's operational issues and support the continued expansion of wind energy production. Many types of renewable energy production already benefit from energy storage technologies. By decoupling the production and delivery of energy from renewable resources, storage technologies can make the generated energy more useful and more valuable—resulting in a better standing in the market for these facilities. To date, the wind power industry has made great strides in enhancing the capability of wind turbines and how they are integrated into the overall power market. Unfortunately, although the direct production cost may now (at certain locations) be competitive with other power generation resources, its effective usage cost is still higher because of inherent qualities of the wind resource. Storage technologies can provide additional flexibility to mitigate these issues. For example, storage can promote system stability to encourage further wind penetration of the market in both small and large power grid systems. Although on small power grids the activity will focus around providing a high-cycle dampening capability (high wind penetration), storage technologies can provide more of a commodity balancing capability for wind on larger power grids, giving system operators a greater level of reliability (through lower variability) for the growing quantity of wind energy. Without addressing these and other negative qualities of the wind resource, these challenges will continue to restrain the development of wind resources and their integration into the market.

Resource Utilization Challenges

Certain qualities of the wind resource itself will continue to hinder future development of wind power generation. In particular, three systemic challenges must be overcome to ensure the widespread adoption of this renewable energy resource:

1. Its noncoincident peak

2. Its nondispatchable nature

3. The stability of the system as wind gains greater market penetration

Wind turbine technology has made great advances in improving the conversion efficiency of this resource, but being competitive on price does not also mean that wind is as useful an energy resource to the system administrator as other more reliable generation technologies. Now that wind-generated power is becoming competitive with other generation technologies, it needs to be made equally useful (for planning purposes) to gain wider market acceptance—especially as its market penetration grows.

Energy storage technologies stand to improve the utilization (and, hence, profitability) of wind projects both now and in the future. It is true that in the near term, there are still many opportunities for wind turbine–only projects to succeed simply by extending the current technology and market strategies. However, even the most profitable of these projects could benefit from the inclusion of a storage component. With the storage component providing a cache for the wind energy, a wind project could approach the sale of its output to the market differently, reducing or even removing the utility-imposed charges that impact its profitability. Storage facilities could even relieve temporary transmission congestion from remote areas with established wind parks that are already threatened by transmission availability issues.

In the long term, these resource utilization challenges will bear directly on the level of market penetration wind power can attain. So far, the existing low level of wind penetration has not been difficult to account for on utility power grids. However, power system operators are hesitant to increase the penetration of such intermittent resources indefinitely. Some way to balance their output must be found, and the cycling of fossil power plants to both demand and wind output variations might not be a workable solution—especially to the owners of the fossil power plants. In fact, small island power grids installing wind turbines already have such problems and are finding storage a solution to many of their needs. Finally, the vast majority of U.S. wind resources are in the West, especially the Great Plains, far away from the major cities and industrial areas. If plans to export wind power to these major demand markets are ever to be considered seriously, it must be done efficiently to reduce the required transmission investment—something already in short supply. Storage technologies could assist in bringing this power to market—a crucial step for reaching many of the various states' renewable energy targets.

Noncoincident peak

The first market challenge in delivering wind energy to the utility grid is that the wind resource is noncoincident with the peak power demand—it simply does not blow the hardest at the best times to produce electricity. For many sites, upward of 67% of the total wind power resource can be outside of the peak demand period (i.e., 9 a.m. to 5 p.m. Monday through Friday), as much of the wind power is only available in the morning and evening. Besides this daily variation, the average wind speed for a particular site also follows a seasonal pattern, with maximum winds occurring (for most locations) in winter and spring, and minimum winds in summer and autumn. These trends in the resource potential present a problem for many utilities because summer is the peak demand period for most of the United States. In fact, for much of the Great Plains (where the majority of domestic wind resources exist), the summer wind power resource can fall anywhere from 20% to 30% below the wind power of winter.

This mismatch of wind resources to power demand has an obvious impact on the profitability of the wind turbine or wind park. Because wholesale electric power prices vary throughout the day, *when* power is sold matters just as much (or more) as *how much* is sold in determining the total value of a wind turbine's output. Unfortunately, unlike other power generation facilities, wind turbines cannot choose when to produce, leaving control over any market strategy the installation may have not in the hands of the owners. Therefore, as the existing strategy is simply to produce all that one can—whenever one can—and hope for the best, any strategy that could maximize higher- over lower-margin production would be well received.

Nondispatchable

The second market challenge in delivering wind energy to the utility power grid is that the wind resource is intermittent—it is unreliable from a scheduling point of view. Not only is the timing variable, but also the intensity is correspondingly variable—leading to uneven power production throughout the day. This variability in the resource combines to leave many sites with an average utilization of only 33%, although siting in exceptional locations or using the latest wind turbine design is pushing this toward 40% or higher. This unpredictability relegates wind energy to a different status than power from conventional power sources—still welcome for a number of reasons, but with definite reservations from a scheduling point of view. In a small power grid environment, for example, this unpredictability requires the system operator to maintain additional dispatchable resources to back up the wind generators, whereas in the broader wholesale power market, system operators account for their additional balancing cost in part through a reduction in the capacity payment to wind developers—often by 80%. Besides losing this capacity payment, unpredictable wind resources do not allow wind generators to maximize their guaranteed delivery of power during peak pricing periods.

. The wind industry has undertaken significant efforts to overcome this dilemma by extending the scope of existing weather forecasts into effectively highly detailed wind speed forecasts. Most of this effort has taken place in Europe, where reliability for hour-ahead wind speed (and hence, potential wind energy) forecasts has reportedly reached a 95% or more accuracy rate, allowing wind resources to participate in short-term energy imbalance and even load-following roles. Unfortunately the day-ahead forecasts continue to fall short of this mark, requiring additional generation resources to be maintained in more traditional standby availability in case of a larger energy imbalance.

System stability

The two previously mentioned market integration challenges for wind resources conspire to produce system stability issues as the magnitude of wind resource penetration reaches approximately 15% of an area's supply. This is not a hard-and-fast rule, as power grid geometry, transmission capability, and responsiveness of existing generation facilities all play a role in determining at what point the variability of the wind power begins to affect the stability of the power grid—especially in areas where transmission constraints exist. Although there is not a significant amount of experience in the United States on this issue, in a number of places in Europe (especially in Denmark), wind energy accounts for 20% or more of the energy production for the power grid. Their experience is that this high of a percentage of wind energy produces unstable conditions on the transmission power grid, akin to (according to some European power grid operators) "driving an articulated lorry with no brakes and no steering." The setup requires additional management adjustment of dispatchable generation resources—increasing production costs and environmental output—in addition to adjustments for a changing load during the day. As wind resources continue to be developed, especially in highly localized remote areas, these problems will become more apparent in the United States over the coming years.

Early evidence in the United States shows that as wind penetration reaches 10% of an area's load, the system operator has charged up to 10% of the value of the electricity for regulation purposes. As the penetration level deepens toward 20%, the additional charges increase only slightly, but shift toward day-ahead energy imbalance costs instead of hour-ahead regulation. Although the charges for this can vary greatly depending on the location, timing, and market penetration level of the wind resource, they are substantial. Studies on the integration cost of wind power have ranged from $1.80 per MWh (Xcel utility study) to $6 per MWh (PacifCorp utility study).[3] Because most potential wind energy power generation sites are located in remote locations where there is generally a limited amount sufficient transmission capacity, a large amount of localized wind energy production in these rural areas can strain existing power transmission systems and potentially limit the transmission of this power to the demand centers.

Finally, besides the daily balancing issues, adding significant wind energy resources to a region's power grid can also affect the regional power flows and, therefore, must be taken into account when accomplishing resource planning or transmission upgrade studies. The problem could even put into question the ability of states to hit their renewable energy penetration targets. For example, California's mandate for 20% (or possibly 33%) of its generation coming from renewables by 2017 may prove problematic, as even existing wind turbines can be underused at times because of transmission constraints. This is of real concern, as illustrated by Ireland; with only 7% of its generation coming from wind power, the country imposed in early 2004 a moratorium on the connection of new wind power because of power grid instability issues. Other states with similarly aggressive goals as California's may also have similar experiences, making it difficult for them to reach their mandates in their renewable portfolio standards.

Remote Power

Renewable energy technologies have long been a favorite for small, remote power applications; adding a storage component to these systems is not simply useful, it is often a necessity to make the setup practical (especially for use after sunlight hours as seen in figure 5–2). The desire to use renewable energy sources for these remote applications is generally based on one of two main drivers: environmental impact or cost. In these off-grid systems, many installations are in environmentally sensitive locations without the benefit of system power, creating a bias (or even mandate) toward using renewable resources over a diesel generator with its accompanying emission and leakage issues. As for cost, connecting the small load to the nearest power grid is also usually impractical because the cost of running a distribution power line to a remote site can quickly rise past any economic validation. This solution has its own environmental impacts, besides the wildly expensive cost per delivered kWh for a small, intermittent load. If self-power is chosen, even the diesel option can be too expensive from a total-system cost perspective, because reliable diesel fuel delivery can become impractical or cost-prohibitive because of the transportation costs, which can easily account for more than the fuel charge.

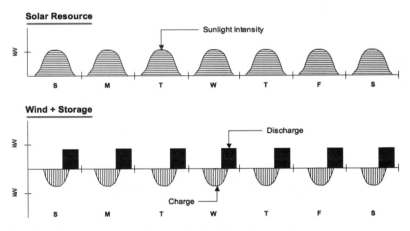

Fig. 5–2. Remote power (Ardour Capital Investments).

Because of the limited power availability (many systems are less than 1 kW in size), these renewable energy power units are generally restricted to supporting remote communication equipment or a limited residential network or other commercial electrical equipment. The renewable technology most often coupled with a storage unit for these applications is photovoltaic (PV), although some small wind turbines are also used in the larger systems. In 2003, nearly 750 MW of solar PV panels were delivered globally, with the market expected to maintain its 30% plus growth for the foreseeable future; of this, roughly 25% goes toward the off-grid market. To minimize operating and maintenance costs, standard lead-acid batteries are most often used to store the power until needed, because their operating capabilities and costs are well understood for these applications, and the duty cycles are generally not substantial. However, most batteries used in these installations were originally designed for the automotive market, so, although many of these systems are of minimal size, they must contain sufficient power electronics to prevent either under- or overcharging that would reduce the operating life of the battery. In general, the solar panel component of these remote power systems makes up approximately 50% of the total system cost, with the storage component often representing approximately 20% of the total system costs. Although it adds to the cost, a storage component does provide economic benefits. With storage, the solar component can be undersized for the required load, thereby reducing the capital costs of the system—possibly by upward of 10%—while expanding the usefulness of the system significantly.

International prospects for these remote power installations far outnumber U.S. domestic opportunities. Many of these projects are concerned with bringing electricity to remote villages where reliable power grid power is not available. In these locations, governments are supportive of these programs because many of these areas may not be served for many years by the expansion of the national power grid. A number of good examples of these projects can be examined through the National Renewable Energy Laboratory (NREL) Village Power Program. Some are discussed in the following sections.[4]

Example—Malaysia.[5] Remote PV systems from BP Solar Malaysia and its joint venture partners—Projass Engineering and Projass Enecorp—are cornerstones of the Malaysian Ministry of Rural Development's (MORD) rural electrification scheme. This program provides electrification to remote communities in the peninsular and the East Malaysia states of Sabah and Sarawak, in one of the largest rural infrastructure projects in the world. These intrinsically simple, stand-alone PV system packages were specifically designed to deliver small-scale AC power in an easy-to-maintain package of solar panels, power electronics, and lead-acid batteries (flooded & gelled). Since 1996, MORD has spent $39 million on this ongoing program to install more than 11,600 PV systems (2,200 kW in total for all systems). These units provide power to individual homes, long-houses, rural clinics, community halls, schools, and churches for uses ranging from basic lighting to vaccine refrigerator-freezers; these systems are now responsible for generating 3,300 MWh per year.

Island Grid

Island grid is simply another name for a small, isolated power grid, usually serving a single town and its accompanying commercial and industrial load. Because of its small size, its load profile is generally much more volatile than larger systems, and it is not uncommon—as is found in many Alaskan isolated grids—for the system's peak demand to be three times the minimum load, and twice the average load. These systems can be found on an actual island, but they also are small power grids in many rural areas where the existing transmission capacity is severely limited, effectively isolating the area from most balancing support of the main power grid. Because of the need to supply the load over such a variable range, these systems' baseloads are generally small, requiring a number of small additional power facilities to balance and stabilize the load as it shifts throughout the day.

The most common power technology used in these systems is a diesel-reciprocating engine. Although dependable, these diesel units require fuel and maintenance, which are often very expensive because of transportation issues (with the transportation charge much larger than the fuel charge). As the units cycle on and off in response to changing load conditions (some only lasting minutes), operating efficiency declines and emissions increase; such activity also raises the maintenance costs and shortens the unit's life. To reduce the wear and tear on the units from multiple starts, many systems impose a minimum run-time on the diesel units, but this strategy then further lowers fuel savings, driving up costs. For example, diesel generator sets (gensets) only partially loaded can use nearly half-again as much fuel per kWh as when run at full power.

To reduce the run-time (and start-ups) on these units, wind power is often sought as an additional, local energy resource. The benefits of using wind energy can be quite high. A number of studies by U.S. government laboratories (NREL, Lawrence Livermore National Laboratory [LLNL], etc.), have shown that adding wind to a diesel-powered local grid can reduce fuel consumption by 40% to 50% and total costs by 30% to 50% for areas with plentiful wind resources.[6] However, because of the small size of these power grids (lack of system inertia in its limited generators, etc.) simply adding wind turbines to small power grids cannot be done haphazardly—a systematic review of the load and potential additional wind turbines must be undertaken to ascertain potential benefits, and to determine what level of wind penetration is best. For many of these power grids, the opportunity exists to have wind resources well in excess of 50% of the peak load. However, if that much wind capacity (or more) is added, steps must be taken to have practical uses when output exceeds demand. This can include space and water heating, as well as simple thermal load dumps (radiators) if no other use is found.

To alleviate the need for such *creative* solutions for the problem of too much low-cost renewable energy (and system optimization), energy storage units should be added to these isolated grids for three basic reasons:

1. Reduction of diesel starts/run-time

2. System stability

3. Increase in the level and value of wind penetration for the system

Because the most variable cost of a small power grid is the diesel operations, the goal is usually aimed at lowering their costs—not just their fuel costs, but the overall usage costs. Probably the greatest single factor in determining the amount of savings a storage component can provide a hybrid system (wind/diesel/storage) is the duration of the storage unit's discharge during normal operation of the system. Storage units can significantly lower the number of power generator start-ups needed to meet short-term requirements for power (fig. 5–3)—essentially providing a ride-through capability for the system. With a larger storage capacity, load leveling becomes possible. In 1997 an evaluation for a wind turbine combined with a battery energy storage facility for a small power grid in Deering, Alaska, was undertaken. The storage unit was used to provide an extended ride-through capability to prevent a peaking diesel unit from running for only a short period of time. In this modeling study, it was shown that the largest savings came from relatively short-term storage; even a 10-minute ride-through capability from the energy storage unit reduced diesel fuel use by 18%, the diesel running time by 19%, and the number of diesel starts by 44%.[7]

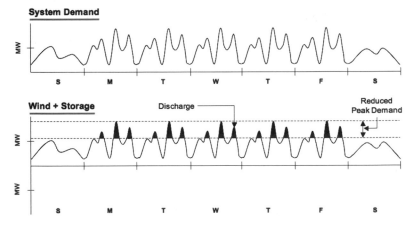

Fig. 5–3. Island grid (Ardour Capital Investments).

Adding a storage component when adding wind power to these small power grids is another way to balance the system and avoid many integration problems. Without a storage component, the system operator must increase the spinning and regulation reserves (incurring a relatively greater cost than on a larger system) to prevent variable wind production from impacting the customer base. As an island power grid is small with little of the deep system inertia found on larger power grids, changes in both power production and system demand have a greater impact than on a major grid. Acting as a *shock absorber* during variations in system stability by providing enhanced frequency regulation and load-following services, these energy storage units have a beneficial impact on the operating costs of these systems. As an additional ready-reserve of power, the unit could also provide spinning or even restoration services to the power grid, enhancing the security of the local utility.

Wind penetration can be very high on these island power systems, easily more than 50% with the right wind resources. The same U.S. government studies that showed that increasing the wind penetration can lower the diesel fuel costs on these systems also showed that adding a storage component can gain an additional 10% to 20% in system cost reductions. Because some of the primary benefits to adding storage surround the role of supplanting diesel starts, storage contributes very little to improved system economics at low wind penetration levels. However, the value of storage to the system increases as the wind penetration increases, as there will be an increasing amount of time that the available wind power exceeds the total system loads. According to one NREL study, at 50% wind penetration, storage can provide 20% greater fuel savings and 20% less diesel run-time than nonstorage wind/diesel systems alone.[8]

Example—Hluleka Nature Reserve, South Africa.[9] Until 2002, the Hluleka Nature Reserve in South Africa relied on two diesel generators for all of its power needs. Unfortunately, it was far from an optimal solution. Billed as a natural retreat, the constant throbbing of the generators spoiled the peace and quiet of this exquisite coastal spot. In addition, although all of the kitchen and bath appliances supplied in each chalet were electrical, transportation and cost issues kept the two generators operating only 10 hours each day. Even with such a reduced load, these units still consumed around 90,000 liters of diesel fuel annually. Although a recognized problem, the reserve needed the flexibility of the diesel engines, because with 12 chalets of guests (plus the management staff), it had to be able to provide for possibly 90 people at any one time.

To improve the quality of the stay for the guests, the country's National Electricity Regulator (NER), the Council for Scientific and Industrial Research (CSIR), Shell Solar, the Department of Minerals and Energy (DME), and the Eastern Cape Provincial Government deployed a stand-alone mini-grid for the reserve, which combined the existing generators with a wind, sun, and storage component to produce the most cost-effective and efficient energy solution. Besides solving the electrification problem, the grid was also designed to fix related issues. Prior to this project, the water supply for the camp was pumped directly from the river into holding dams without any purification process, and telecommunication in the area was, at best, problematic. The mini-grid was designed to not only provide sufficient electrical power for the reserve, but also to provide sufficient power for enhanced water purification and communication systems.

The resulting hybrid system makes use of a 56-kW solar array, two 2.5-kW wind turbines, a battery storage unit, solar water heaters, and liquid petroleum gas. On sunny and windy days, the power from the generators is fed directly to the camp and to the battery storage unit, which can supply five days of reserve electricity in case of no wind or sun. As a backup, one of the existing diesel generators is being retained. Because upgrading the telecommunication equipment was part of the project, an Internet connection allows the batteries, wind chargers, solar panels, and power consumption to be remotely monitored 24 hours a day by Shell Solar.

Example—Metlakatla, AK.[10] One well-known example of how a storage facility can improve an island power grid is a 1997 DOE-sponsored (through the Energy Storage Systems program at Sandia National Laboratory) project in Metlakatla, AK. Here, the full cost of the $1.5-million energy storage facility was recovered within three years because the storage unit improved overall power quality and reduced the operating costs of the existing generating assets. Metlakatla is an 800-resident island community in southern Alaska (common in this area of the state), with an economy based on fishing (commercial cold storage) and a lumber mill. Taking advantage of plentiful water resources, the local utility powered its local grid with two reservoirs with four turbines that produced 4.9 MW of hydroelectric power. In 1986, a chipper was added to the sawmill—causing the hydroelectric dam to struggle against brownouts, overvoltage, and frequency fluctuations when responding to load swings of as high as 900 kW. Because the hydropower dam's response time of 10 seconds was too slow to follow the rapid swings in demand from the sawmill, which could shift in as little as 1/20th of a second, the village often suffered brownouts and sometimes blackouts. A 3.3-MW diesel system was installed in 1987 to alleviate this problem; the unit typically operated at 20% to handle load swings and provided 55% of the system's power. With this diesel genset, generating capacity grew to be twice the average system load; unfortunately, continued inadequate spinning reserves and erratic system frequency combined to produce continued poor power quality for the utility's customers. It was decided that a more workable long-term solution was needed, because even this poor level of service cost on average more than $400,000 for diesel fuel, a nontrivial amount of the community's scant resources.

Sandia's Energy Storage Systems program developed a lead-acid battery (batteries from GNB Industrial Power, a division of Exide Technologies) energy storage system that was installed in 1997 (with a 20-year expected operating life). Surprisingly, the system matched generation needs very well, almost totally replacing the diesel unit in normal operation. (The diesel system was then operated primarily to charge the battery when the hydroelectric system underwent maintenance, or during drought conditions.) To minimize operating costs, the energy storage system can operate autonomously for normal charge/discharge, standby, and disconnect/synchronization for blackstart.

There were both direct financial and nonfinancial benefits to using the system. One financial benefit was that the storage facility greatly reduced and nearly eliminated the annual transportation of 500,000 gallons of diesel fuel (at a cost of $400,000 per year) from the mainland by ferry and pipeline. Also, each of these diesel shipments also required an initial outlay of $100,000 for handling costs—a serious burden for such a small community. In addition to the fuel costs, interim diesel unit maintenance cost $150,000 every three years, with major overhauls of $250,000 required every six years. By reducing the constant starts and extended operating time, the storage unit has increased the time between required overhauls.

Nonfinancial benefits were substantial for the community. The energy storage facility greatly reduced the environmental risks to the community because diesel emissions and noise were greatly reduced (or eliminated). Productivity improvements were evident as enhanced power availability and quality became the norm for the community's commercial fishing operation (the cold-storage facilities) and sawmill. Outages from the earlier power disruptions were virtually eliminated, along with the financial losses for the corresponding product spoilage.

Example—Dogo Island, Japan.[11] The installation on Dogo Island, a small island just off the coast of Japan, is an excellent example of how a flywheel energy storage system can provide the stabilizing capability lacking on many island power grids, which had inhibited their use of the readily available—but highly variable—wind energy around them. In 2003, Fuji Electric installed a 200-kW Kinetic Energy Storage System (KESS) from Urenco Power Technology (UPT) in conjunction with a 3 x 600-kW installation of DeWind D4 wind turbines to evaluate how wind generators can be a viable source of power on remote islands with weak links to the mainland power grid by smoothing their irregular power output. Through incorporating the flywheel-based energy storage unit into the installation of the wind turbines, Fuji Electric sought to fulfill four goals; they wanted to

1. Stabilize the frequency variations stemming from the turbines

2. Capture excess energy from short-term wind gusts

3. Optimize the operation of (or eliminate the need for) diesel generators on the island

4. Eliminate the need for additional spinning reserve because of the introduction of the wind turbines

Results to date have been promising; by acting as both a dynamic sink and source of energy, the UPT KESS improved the island's power grid efficiency and increased the penetration rate of the wind turbines. The flywheel unit's ability to provide a stabilizing capability to the highly variable wind turbine power was found to be essential in allowing Fuji Electric to connect the wind turbines to the island's relatively weak electrical transmission system. Because of this successful outcome, Fuji Electric is now looking for further deployment opportunities of the UPT KESS technology to provide truly reliable wind-generated energy as a viable supply alternative in other locations.

Example—King Island, Australia.[12] A more recent example of a storage facility assisting an isolated power grid to use wind energy can be found on King Island, a remote island located off the southern coast of Australia. The local utility, Hydro Tasmania, installed five wind turbines (ranging from 250 kW to 850 kW) over time to supplement and hopefully supplant the power from four 1.5-MW diesel generators. As with many isolated power grids, the demand profile was volatile, with the peak power levels often reaching 80% above the average system load. Unfortunately, the wind resources in the area are also extremely variable, with much of the wind energy not coincident with the peak power demand of the local residents. Therefore, the desire to add a renewable energy resource to the island's power mix actually led to a situation where the existing diesel unit performance declined from constantly cycling to balancing the changing consumer demand and variable wind turbine output.

Hydro Tasmania decided to add an energy storage unit both to improve the balancing of resources with demand and to add another component of system stability. The storage unit selected for installation in 2003 was a 200-kW (800 kWh) flow battery from Pinnacle VRB Ltd (a subsidiary of VRB Power Systems). Factors contributing to Hydro Tasmania's decision to choose the flow battery were the unit's rapid charging capability (1-to-1 charge/discharge) and its round-trip efficiency of 77%. To assist in the deployment of the unit, the Australia's Renewable Energy Commercialization Program provided a grant of $700,000AU.

Since the vanadium redox battery energy storage system (VRB-ESS) unit began operating, it has provided three main benefits as it lowered operation costs for the utility and reduced the system emissions by 47%. First, it allowed load shifting of the off-peak wind energy to increase the contribution of the existing wind turbines to the island's power needs. Second, the storage unit improved operation of the existing diesel engines. With the additional stored energy as a resource (and the ability to absorb temporary excess wind power), the diesel units could be dispatched more economically, resulting in both a reduced number of starts and reduced run-time for the units. This provided both reduced fuel usage and lower emissions of diesel exhaust. Overall, as wind energy increased in its share of supplying the island's power, the total diesel run-time had been reduced 1,100 hours per year, reducing fuel usage by more than 1.5 million liters/year. Finally, the storage unit improved stability through better frequency regulation, load-following, and enhanced voltage control for the system as the variable wind resources were used to supply more power to the island customers.

Grid Connected

Although wind power is quickly becoming a real solution for remote-power and island-grid environments, the real market for wind power development lies with large grid-connected wind parks.

To improve their unit competitiveness, the focus of much wind turbine development has been to reach cost parity with natural gas–fired combined-cycle units. For large modern wind turbines, that goal has been reached; however, being competitive on price does not mean that wind is as useful an energy resource to the system administrator. Currently, to ensure a wind turbine's competitive position not only requires that the turbine be placed in an advantageous wind resource area, but also that the demand will be there for this power (rate and quantity), and that demand will be within easy transmission distance. For wind that only blows hard at midnight in a rural setting, the only competitors a wind turbine using this resource will find are baseload coal units idling at low power—a competitor no power generator relishes. Although not a panacea, energy storage technologies can provide assistance tailoring the wind turbine's operating strategy for different markets by decoupling the production of wind power from its immediate sale.

There are three basic strategies for coupling storage technologies with large wind projects: dispatchable wind, capacity firming, and baseload wind. Interest in storage does exist in the wind community, but what wind developers want most from any type of storage technology is something that is cheap, responsive, and reliable. As large-scale wind parks are relatively new compared to other power generation options, storage will be introduced slowly, building from experience in the island grid arena to increase flexibility and alleviate existing integration costs. In the long term, storage can be more systematically integrated into wind projects as developers better understand how to leverage a storage technology's ability to reduce the risk associated with wind projects, enabling a greater level of market penetration for wind overall.

Dispatchable wind

An early strategy to improve the economics of wind power was simply to capture all of the power from a wind turbine and sell it as a single block into the market. By using a storage facility to totally decouple the production and delivery of the power, the wind energy could be made dispatchable and sold during peak demand periods as firm power—transforming the low-value, unscheduled output of the wind turbine into high-value, on-peak *green* power and ensuring the maximum revenue for the wind park. Beyond wholesale power sales, the strategy held out the potential for the facility to sell capacity to ISOs for ancillary services such as contingency services or restoration services.

Storing all of the power for a sale as a single block of power (fig. 5–4) was designed to work around some of the less useful aspects of wind power production. First, because the power delivery is firm, there would be no capacity payment reduction nor any additional discounting by the utility to make up for added ancillary service costs. Second, this strategy allows for capturing the highest peak power prices, especially in the summer months when output (especially in the Midwest) is noticeably reduced at the time power is most desired. With the ability to obtain high market prices, this strategy also holds out the potential to develop marginal resources not normally considered cost-effective. Interestingly, although the dedicated transmission power lines are totally unused during the majority of the day when the storage facility is charging, the overall usage of these assets would essentially be the same as if the wind turbines delivered their power to the grid as it was produced (if the same size power line was used).

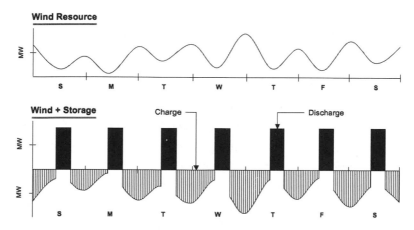

Fig. 5–4. Dispatchable wind (Ardour Capital Investments).

Although straightforward, this strategy has only seen limited interest recently, and it is doubtful that it will garner much future development. The reason for this is that the added value of selling a firm block of power into the market cannot usually make up for the loss in market flexibility and added cost of the storage component. Part of this stems from the recent advances in wind turbine technology, which have reduced much of the lower efficiency issues of earlier wind turbine technology that this strategy planned to correct. Also, selling capacity for contingency reserves or restoration services would be problematic as the full power capacity for these ancillary services would only be available daily for a few hours just after the storage facility was fully charged but before the unit would discharge. Also, selling any of these services would preclude selling the bulk power into the market. Finally, this strategy relies on having unfettered access to sufficient transmission capacity for anywhere from four to eight hours per day. However, depending on the regional transmission rules, it is possible that rural areas would have insufficient transmission capacity (contractual and/or physical) during essential hours, constraining the facility's output off the grid, and thus losing some or all possible revenue for that day. If for any reason the output could not be delivered to market, penalties in the contract (for firm power) would also significantly reduce the profitability of the project.

Example—Alta Mesa, CA.[13] The Alta Mesa pumped storage project (TenderLand Power Company), which was proposed in the late 1990s for construction in the San Gorgonio Pass near Palm Springs, California, is a good example of an early project designed to store off-peak wind energy and sell it into the market during peak demand. The 70-MW pumped storage facility was to be associated with a 54-MW wind park and have a round-trip energy storage efficiency higher than 70%. By capturing the wind energy produced during off-peak hours, 420 MWh could be delivered during on-peak hours to Southern California Edison's transmission system. The facility was to consist of an upper reservoir on the Alta Mesa and a pair of large concrete tanks at the bottom of the hill with a hydraulic head of 1,250 feet and a storage capacity of 113 million gallons in a closed-loop system. Part of the initial interest in using storage on this wind power project was that at this location, the majority of the power generated is produced during off-peak periods, limiting the economics of the wind-only development. By incorporating a pumped-hydro storage facility, the combined project could sell six hours of firm green power during peak demand (and pricing) periods, increasing the value of the intermittent wind energy.

Capacity firming

If energy storage technologies are to play a significant role in conjunction with wind power, it will be through firming the delivery of wind power from grid-integrated wind parks. Rather than cycling all of the output from the wind turbines through the storage facility, the capacity-firming strategy focuses on providing sufficient support to the output of the wind park to ensure maximum on-peak energy sales (fig. 5–5). Many of the current negative aspects of selling wind power into the market—lack of capacity payments, additional ancillary services, and energy reserve support requirements—are then reduced, improving the actual economics of the wind park. Although this strategy cannot provide full power-dispatchable green energy resources for the grid in the same way as dispatchable wind, it can significantly improve the project's ability to provide some level of firm power from the wind park. In this way, the wind developers' customers—system operators—can begin to view wind in a far more positive light, especially as the amount of wind energy delivered to the grid continues to climb.

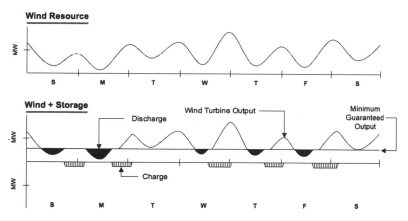

Fig. 5–5. Wind capacity firming (Ardour Capital Investments).

Determining the needed power rating for a storage unit to support a wind park for this strategy requires knowing and understanding such issues as the size of the wind park, the variability of the local wind resource, the local power transmission capability, and the average load profile of that local power system. With a volatile wind resource site, it is not uncommon for the output of the wind park to change suddenly, possibly exceeding the capability of the transmission system to absorb the power. Surprisingly, the required power rating of the storage unit only needs to be a small portion of the rating of the wind park to be useful. This is because the power rating needs to only be as large as the expected variations in the output of the wind park from some long-term median desired output level. This often requires sizing the power rating of the unit at only 20% of the size of the wind park.

More important to the strategy of the ensuing combined wind and storage combination is the determination of how much energy storage (MWh) is required. For areas of relatively constant wind speeds, high cycling energy storage facilities with only a small storage capacity may be useful. Akin to the frequency regulation service provided on small island grids which are constantly absorbing and discharging a larger amount of this excess energy, they can improve the delivery of a more stable power output from the facility, benefiting the power flow on the grid. This would especially benefit areas with weak transmission systems. On a more common basis and for larger wind projects, wind energy during off-peak periods or during any time when the transmission of the output is constrained could be stored and then used later to supplement existing wind generation during on-peak periods to ensure maximum energy sales of firm green power. The marginal cost of adding additional storage capacity for most energy storage technologies is relatively cheap (compared to the cost of the entire system in general), opening up additional strategies. This additional power could also be released during low- or zero-output periods of the wind park to provide minimal energy delivery. The amount of energy storage capacity would then determine the length of time for which output could be supported without wind generation and help with possible reserve requirements beginning to face wind projects. Currently, the only way to increase the capacity factor of a wind park is to oversize the number of wind turbines for the transmission system (with the extra wind turbines providing the reserve capacity), but developers are loathe to follow this strategy to provide these contractual reserve levels.

Example—McCamey, TX.[14] Texas ranks second amongst the 50 states for its wind potential, and McCamey, Texas, has been the center of much development activity because of its wind resource potential. Unfortunately because of its remoteness in West Texas, developers are building more wind generation than the transmission grid can easily handle, causing congestion problems. Although plans exist for additional transmission upgrades, new wind park development is expected to match or even surpass these upgrades. To alleviate this near-term transmission constraint problem and provide room for additional wind generation in the area, Texas's State Energy Conservation Office (SECO) commissioned a study (led by the Colorado River Authority) to determine what benefits a large-scale energy storage facility would have for transmission. To support this expected continued mismatch between wind power and transmission capacity, SECO chose a CAES facility with 400-MW compression, 270-MW generation, and an extremely large 10,000 MWh storage capacity (at full power, a 25-hour capacity for compression, and a 37-hour capacity for power generation). The facility would be used to store power during periods when congestion on the transmission lines constrained the growing wind resources.

Unfortunately, because of the wind pattern in the area—there are times when the wind blows strongly and continuously for days at a time— the modeling of the project showed that the CAES facility could become fully charged and unable to provide additional compression even when it would still be required. As providing total reduction of transmission curtailment was the single desire for the study, it was found that the CAES facility could not alleviate 100% of the transmission constraint; thus it could not be evaluated as a substitute for transmission line capacity. However, extending the evaluation of the CAES facility past a purely transmission replacement role, it was found that it could provide several benefits:

- The facility would provide more wind generation (up to 400 MW) with only minimal curtailment, and better capacity firming of the wind parks for the area—providing assurance that far more power could be delivered during peak demand periods.

- Combining wind with storage would ensure that wind could claim credit for operating reserves (equal to the amount of CAES generation). Although this was not a significant payment, having this capability added to the total value of the facility (the value of any facility stems from not just one revenue stream, but many), and it provided additional firm capacity for the system operator to call on. This will be needed in the near future, especially as the amount of wind generation continues to grow.

- By operating the CAES facility to support wind generation in the area, wind energy curtailment reduction totaling over 600 GWh annually was achieved (compared to the area without the storage facility). This resulted in several million dollars of profit annually above and beyond what would be required for a positive return on the CAES facility investment. Because of this outcome, work continues on siting a CAES facility for this role in the area.

Baseload wind

The third market strategy—baseload wind—incorporates storage most closely with the delivery of wind power to the power grid. It is an extension of the capacity-firming concept and is envisioned for the largest scale wind storage projects, with most facilities reaching hundreds of MWs (or more) in size. Although most wind turbine technology and development has (rightly) focused on the production of low-cost wind power for direct sale, this concept (long championed by Alfred Cavallo) focuses on obtaining the maximum wind energy production from a particular site, and then using the storage facility to optimize the delivery of power to the market competitively (fig. 5–6). This then creates a choice of different wind turbine designs and maximizes their site placement because there is no concern with their output being constrained from delivery during peak periods. By balancing and optimizing the wind turbines, storage component, and transmission of the power, this strategy is designed to provide wind-derived power to markets with the same capacity and dispatch capability as a mid-merit or even a baseload power plant—especially over long distances.

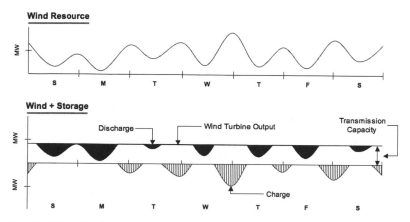

Fig. 5–6. Baseload wind (Ardour Capital Investments).

During normal operation, the output of the combined facility would provide power primarily from the wind turbines, supplemented by energy from the storage unit. Wind energy could be stored both during off-peak periods (evenings and weekends) and during periods when the output of the outsized wind park could be constrained from delivering power. Because of the storage capacity required and the flexibility in siting, the storage technology most envisioned for this would be CAES. In this design, both the size and the capacity of the storage facility are somewhat flexible, with the power rating of the storage unit sized to accept the variations in output from the wind park and the facility's energy storage capability (10s of hours at a minimum). In this way, the storage facility is sized in relation to the amount of capacity support envisioned by the project—with a larger storage component translating into a higher overall reliable delivery from the facility. With this design, instead of an average wind park's roughly 33% utilization rate, the mix of wind power supplemented with CAES output could provide double that, or more. With the significant storage capability of the CAES facility, even seasonal storage of the wind energy would be possible.

Besides providing system operators a means to incorporate wind power into the market in a reliable manner, a baseload wind project could also be the basis for transmitting wind power over long distances. Much of the U.S. wind potential is simply too far away from demand centers to ever be built out. However, the reliable output of a large baseload wind facility could provide long-distance transmission through a dedicated long-distance high voltage power line. Because delivery of the wind park's power would be regulated by the storage facility, a much higher utilization of the capital-intensive transmission system could be achieved—making the project economical, unlike the current setup. Large, long-distance high-voltage direct current (HVDC) transmission of site-specific power exists, whether for hydropower from the Pacific Northwest to California or lignite coal power from the upper Midwest to Minnesota. Although siting large power lines—for any reason—is extremely difficult, it is not out of the question that more such lines could be built for wind power, especially if those involved are serious about significantly increasing the penetration level of wind power into the overall U.S. power market.

Example—Iowa Stored Energy Project (ISEP).[15] As a first step toward a large-scale baseload wind facility, the Iowa Association of Municipal Utilities (IAMU) is developing a utility-scale wind park that will incorporate a large-scale energy storage facility near Eagle Grove, Iowa. The Iowa Stored Energy Project (ISEP) is a $200-million integrated power facility, composed of an 84-MW wind park connected to a 200-MW CAES facility. The compressed air for the CAES plant will be stored in an underground aquifer located 1,200 feet below the surface. This would be the first use of an aquifer for a CAES facility, as both of the existing CAES facilities use underground salt caverns. The aquifer has already been used for natural gas storage; in fact, because of the structure of the underground rock, both air and natural gas storage are envisioned for the site (held in different strata); the stored natural gas is to serve both as a fuel source for the CAES plant and as a seasonal natural gas storage facility for IAMU members. Storing natural gas on-site for the power facility will also alleviate a potential pipeline capacity shortage during peak operating periods. The potential storage capacity of the aquifer is huge, and is estimated to be able to store anywhere from 8 billion to 30 billion cubic feet of compressed air at 500 psi (the Macintosh, Alabama, facility can hold 19 million cubic feet at 1,100 psi).

The IAMU members have decided to proceed with the ISEP because the project presents a unique way to generate additional power needed to meet annual load growth, while capitalizing on a number of inherent strengths. Iowa has tremendous wind resources, and increasing numbers of these projects are finally finding acceptance in the regional power market. CAES was seen by the IAMU members as a means to capitalize on this growing local renewable energy resource base and provide the flexible and reliable mid-merit capability needed for customer supply. By using renewable energy and highly efficient natural gas power generation, the ISEP is also designed to reduce the carbon emissions of the IAMU utilities even as their power needs grow.

The ISEP will operate as an intermediate load power plant, producing a mix of wind and power from the CAES facility 12 to 16 hours per day, 5 days per week. During off-peak periods, the ISEP will compress and then store air in the underground aquifer using power from both the wind park and off-peak system power (important in the event of insufficient wind power). Overall, the facility is expected to maintain a utilization rate of 50%, with wind power responsible for 33% of the total output. Combining the competitive economics of both wind turbines and the CAES facility, the ISEP project is expected to be cost-competitive with other intermediate power plants in the region. Gradually, with the addition of more wind turbine generators and air compressors, the facility could evolve into a baseload facility in both scale and capability. One option for expansion uses 400 MW of wind turbine capacity with the 200-MW CAES power train, which would allow the facility to operate with a utilization rate of 75%—similar to surrounding coal-fired baseload facilities.

References

1. Energy Information Agency. 2003. *Renewable energy annual—2002*. Washington, DC: U.S. Department of Energy.

2. Pacific Northwest Laboratory. 1991. *An assessment of the available windy land area and wind energy potential in the contiguous United States*. Richland WA: Pacific Northwest Laboratory.

3. Parsons, B., M. Milligan, B. Zavadil, D. Brooks, B. Kirby, K. Dragoon, and J. Caldwell. 2003. *Grid impacts of wind power: A summary of recent studies in the United States* (NREL/CP-500-34318). Golden, CO: National Renewable Energy Laboratory.

4. National Energy Renewable Laboratory. Renewables for sustainable village power. http://www.nrel.gov/villagepower/index.html (accessed March 26, 2005).

5. BP Solar. 2001. *Mord Project—Phase 1&2 (Application Type, Rural Infrastructure)*. brochure. Fredrick, MD: BP Solar.

6. Isherwood, W., R. Smith, S. Aceves, G. Berry, W. Clark, R. Johnson, D. Das, D. Goering, and R. Seifert. 1997. *Remote power systems with advanced storage technologies for remote Alaskan villages* (UCRL-ID-129289). Livermore, CA: Lawrence Livermore National Laboratory.

7. Shirazi, M., and S. Drouilhet. 1997. *An analysis of the performance benefits of short-term energy storage in wind diesel hybrid power systems* (NREL/CP-440-22108). Golden CO: National Renewable Energy Laboratory. (Also presented at the ASME Wind Energy Symposium, Reno, NV, Jan. 6–9, 1997.)

8. Ibid.

9. *South Africa: Sustainable energy for Hluleka Nature Reserve*. 2002. Shell Solar Project Brief. Amsterdam, The Netherlands: Shell Solar.

10. Butler, P. 1998. *Energy 100 awards—Metlakatla Energy Storage System*. Sandia, NM: Sandia National Laboratories.

11. Urenco Power Technologies. Case study, Fuji Electric, wind power study. http://www.uptenergy.com (accessed March 26, 2005).

12. Hennessey, T. *The multiple benefits of integrating the VRB-ESS with wind energy producers—a case study in MWH applications*. Paper presented at the American Wind Energy Association (AWEA) Conference, Chicago, April 2004.

13. Chippas, S. J., P. R. Blaszczyk, and M. G. Jones. 1999. Wind and water, the Alta Mesa project. *Renewable Energy World*. 2 (2).

14. Study of electric transmission in conjunction with energy storage technology. August 2003. Prepared by Lower Colorado River Authority for the Texas State Energy Conservation Office, Austin, TX.

15. Wind, T. 2002. Compressed air energy storage in Iowa. Wind Energy Consulting. Report prepared for the Iowa Association of Municipal Utilities, Ankeny, IA.

6 OUR NEW ENERGY FUTURE

Energy storage technologies have garnered interest throughout the power industry because of the growing evidence of their expanding capabilities. For this interest to continue, however, these technologies require a road map for their future development to better align themselves with the current needs of market participants and to be capable of adapting to a changing environment. This road map cannot dictate which customers will use storage technologies and when, but it should align the capabilities of storage technologies with interested parties who will drive this technological development and deployment. To understand how these energy storage technologies will work in the market, much can be learned from the time and effort expended to reach this point on the development curve; and storage technologies have played a valuable role in reaching a variety of customer goals.

The Road to Here

Many different groups have supported the development of energy storage technologies. Through previous research and demonstration projects, these groups have been responsible for the commercialization of many of the existing available energy storage technologies. Current programs will continue to be instrumental in commercializing the most recently developed storage technologies now just exiting the lab and entering pilot-phase projects.

U.S. federal government

The U.S. federal government has long supported energy storage technologies because of their promising capabilities. Since the establishment of the industry, public policy makers have attempted to protect customers from volatile prices or unfair trade practices while improving operation conditions for businesses. To this end, they evaluate the changing economic conditions of the industry to craft public policies and support promising technologies that will promote their long-term goals. Energy storage technology is one such promising technology.

Capabilities of energy storage technologies—improving the efficiency, reliability, and security of the electric power industry—are all goals embedded within the principles of the U.S. Department of Energy's National Energy Plan. This and other recent policy documents reveal renewed efforts to craft an alternative market structure, where self-correcting forces solve today's (and tomorrow's) challenges where former command-and-control measures have not succeeded. To promote the development and introduction of new technologies like energy storage technologies, the U.S. Department of Energy (DOE) is following the success of such environmental regulations as the SO_2 cap-and-trade system. Rather than planning exactly how the goals of SO_2 reductions would be reached, the government helped industry develop a number of new technologies and strategies for market participants to choose from and let the market sort out the most cost-efficient means of reaching the goal. The government understands that using new technology within the marketplace provides the means to reach these goals; hence, it looks to innovative and flexible technologies such as energy storage.

The primary area of support for large-scale energy storage technology research has been from the DOE, but other groups with the federal government have seen the promise of these technologies and initiated their own related programs, including the Department of Defense and the Department of Transportation. In fact, the U.S. government has supported large-scale research on energy storage technologies since the mid-1970s; even prior to the formation of the DOE, the Energy Research and Development Administration (ERDA) was evaluating applications for energy storage technologies. By supporting the basic research and proof-of-concept installations at both the ERDA and the DOE, these federal programs have supported the eventual commercialization of these technologies by the private sector.

Now the DOE channels its support for energy storage technologies through the Energy Storage Systems (ESS) Program. The ESS program endeavors to develop advanced energy storage systems in partnership with industry to minimize costs incurred from power quality and reliability problems, increase technology choices in deregulated, competitive electricity markets, and increase the value of renewable and distributed resources. The increased emphasis and visibility that the DOE is placing on energy storage technologies are evident by the ESS program's inclusion into the recently formed Office of Electricity and Energy Assurance, which formed to coordinate the modernization of the electric power grid.

The ESS program provides a framework for promoting storage technology in the market. By identifying customer requirements and analyzing operations experience of field systems, the program can better target engineering research for storage system component research and integration. Much ESS program research is conducted through Sandia National Laboratories (Sandia, New Mexico), which focuses on consumer- and distribution-side systems. The ESS program also works closely with industry partners, often in a cost-sharing arrangement. This allows the federal government to support research programs at manufacturers with imminent promise that lack the financial capability to finish prototypes of new technology. The DOE also cooperates with major utilities to fund technology proof-of-concepts and validation projects. Showcasing a successful integration and operation of the technological platform is essential to allay customer reticence toward being the first user of any technology. Besides proving the facility actually works, operational data can be gathered under real-world conditions to refine the most profitable market application.

State energy programs

Recognizing the potential benefits of energy storage technologies, a number of state governments have developed programs to evaluate and promote the adoption of commercially ready technologies in real-world applications. These state governments have seen the potential for energy storage technologies either in the marketplace or in development by the federal government, but they have structured the programs to focus on near-term issues facing their particular state economies, such as mandates for renewable energy portfolios or protecting and enhancing the environment for both businesses and electric infrastructure. The hope is that this will help new and existing firms with their power quality issues and also new technology firms that will develop new storage technologies within the state. Two of the most prominent of these programs are found in California and New York.

California. The first example of a state-sponsored energy storage technology program is found in California, where the state government has turned to energy storage technologies to help maintain the state's power system reliability while it tackles a number of challenges stemming from the effect of deregulation on the state's continued load growth. The California Energy Commission (CEC), in conjunction with the ESS program, launched a program in 2001 to provide matching funds for projects to deploy market-ready storage technologies that can provide economic value within three to five years. The goals of the projects are: to improve the operation of transmission and distribution grids by alleviating congestion or providing voltage and frequency regulation support; to support the dispatchability of renewable resources; and to help with peak load reduction and load management. The three winning technologies were announced in 2003 and, once installed, will be monitored for three years, with the units operating for a minimum of 18 months. These projects will receive nearly $4 million from CEC (through a utility CEC surcharge), and the participating firms will fund an additional $4.6 million through cost sharing.

New York. Another example of a state-sponsored energy storage technology program is found in New York, where the state government, through the New York State Energy Research and Development Authority (NYSERDA), initiated a cost-sharing program in 2004 to demonstrate the practical use of emerging energy storage technologies, develop energy storage technologies for commercial production within the state, and perform feasibility studies for the use of energy storage technologies in a number of commercial roles across the state. Similar in structure to the CEC program, the ESS program will assist with technology evaluation and support of the cost-sharing program. The two winning technologies were announced in 2004 and will receive $3.5 million of support from government sources. Once installed, these projects will be monitored through their year-long operational period. Goals of the program include electric service reliability improvement, grid voltage support, transmission and distribution upgrade deferral, time-of-use energy cost management, and renewable energy generating sources capacity firming. This project follows on a series of other NYSERDA projects to enhance the electric power service within the state.

International programs

Many other governments have seen the potential of these technologies to solve unique and intractable challenges and have undertaken research programs (collectively and independently) to develop energy storage technologies. Just as in the United States, these other public policy makers see energy storage technologies as a means to protect customers from volatile prices or unfair trade practices while improving the conditions for business operation. At the same time, they are establishing a base of advanced technology manufacturing— and potentially lucrative export capability—in fields that include power electronics, electrochemistry, and others.

International Energy Agency. Acting as a technical clearinghouse and driving agent, the International Energy Agency's (IEA) Efficient Energy End-Use Technologies program contains 14 different Implementing Agreements (IAs), or areas of research. One of these programs focuses on energy storage. Active participants in the Energy Conservation through Energy Storage program include Belgium, Canada, Finland, Germany, Japan, the Netherlands, Sweden, Turkey, the United Kingdom, and the United States. Although the interest in these technologies is widespread, the interest in electrical energy storage technologies is being driven by four main issues:

1. Widespread reforms in electric utility regulations

2. An increasing reliance on electricity in national economies

3. Prevalent adoption of renewable energy sources

4. Attempts to reduce the environmental impact of power production

Energy storage technologies have garnered such interest because of the rapid rate of improvement in the basic technology, along with anticipated cost reduction of each unit once production commences.

Although much of this research has been focused on thermal energy storage, the IEA sponsored its first specific research program (Annex) into electricity storage with the IEA Implementing Agreement on Energy Conservation through Energy Storage (Annex 9)[1] and a follow-up research program that has been proposed but not yet acted on (Annex 15).[2] This next research program is designed to evaluate storage's capability assisting distributed generation assets to realize their full potential in reducing atmospheric emissions and mitigating climate change.

Governmental programs. Besides the support for the collective research and development (R&D) approach through the IEA, many other governments have established individual domestic research programs into energy storage technologies. Besides supporting the IEA research program, countries such as Japan, South Africa, Israel, and Australia have developed a network of domestic energy storage technology R&D programs.

The goals of these countries are threefold. First, just as in the United States, the governments of these countries see energy storage technologies providing a means to improve the production and efficient use of domestic energy sources. Second, these governments are interested in developing a technological base of innovative energy technology manufacturers with export potential. Finally, the availability of capable energy storage technologies is crucial because they help provide the clean and reliable power quality necessary to promote higher-valued industry and commerce across many industries.

Industry programs

Besides governmental support, many types of firms active in the industry have played an important role promoting and supporting the development of energy storage technologies. In particular, the Electric Power Research Institute (EPRI) has long promoted these technologies, acting as a driver and clearinghouse for technical information.

EPRI has been instrumental in many of the first pilot projects of many of the different storage technologies as they sought to show their value in a number of different applications. EPRI was developed in 1973 (through taking over previous R&D research conducted by the Edison Electric Institute [EEI] and the Electric Research Council) as an industry-led organization to conduct electricity-related research and development programs. One of its latest, the Electricity Technology Roadmap Initiative, represents a collective vision of the opportunities for electricity to serve society in the 21st century through advances in science and technology. In the technology road map, energy storage technologies are highlighted as enabling technologies that strengthen the power delivery infrastructure, foster a revolution in consumer services, and boost productivity and prosperity. As part of its ongoing support, EPRI launched a research program into energy storage technologies and their ability to assist with ongoing transmission and distribution concerns. One of the first stages in this continuing effort was the publication of a detailed technical report on the impact energy storage technologies can have on the transmission system entitled, *EPRI-DOE Handbook of Energy Storage for Transmission and Distribution Applications.*[3]

Outside of the United States, other similar industry groups have supported research into energy storage technologies and their applications over the years. In Japan, the Central Research Institute of Electric Power Industry (CRIEPI) was established in 1951 as a comprehensive research organization for the electric power industry, with a mission to conduct fundamental, pioneering research on the electric industry. In each of the three central objectives of the group (cost reduction and ensuring reliability, creation of integrated energy services, and harmonizing energy and environmental priorities), research into energy storage technologies has been conducted to provide benefits to those in the industry.

Besides these industry-wide groups, individual utilities across the globe have provided key support through their proprietary programs to evaluate a variety of storage technologies across a number of applications.

Utilities such as American Electric Power, Detroit Edison, Entergy, Eskom, RWE, and Tokyo Electric Power Company (among many others) have supported energy storage technologies and their promise as solutions to some of the more intractable problems not currently being met by other technologies. The programs underway promise significant savings and additional flexibility in system operations. In previous years, this support has had far-ranging impacts, so the expectation for today's programs is significant. Throughout the 1960s and 1970s, many U.S. utilities reacted to the rapidly rising cost of fuels by searching for a means to arbitrage some of their off-peak baseload power to daytime hours. They began building pumped-hydro units, and pumped-hydro units expanded their share of the U.S. peak capacity dramatically. As these technologies went into service, additional applications were discovered that increased the value of these installations even more. Unfortunately, as the market changed, the capability of these utilities to build additional similar units reduced—hence the interest in other large-scale energy storage technologies that have a smaller environmental footprint. Without any renewed support for large-scale storage projects, the proportion of these facilities in the overall generation mix has fallen, reducing their beneficial impacts (fig. 6–1).

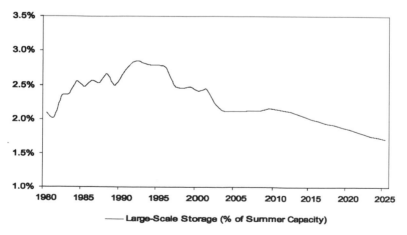

Fig. 6–1. U.S. large-scale energy storage penetration declines (Ardour Capital Investments).

User's Goals for Storage Technologies

Storage technologies have been gaining visibility and interest with the R&D community as well as government policy makers, utilities, and ISO officials. For storage technologies to penetrate different segments of the market and be more widely used, however, their capabilities must continue to align with the goals of the individual players in the market. This is an important aspect for any technology type, but it is particularly vital to technologies such as energy storage, where the focus is on enabling existing operations and strategies to function more efficiently and reliably in a volatile environment such as the electric power industry. To date, energy storage technologies have been able to provide support for the changing goals of market participants, and their flexible operating capability ensures continued consideration as a solution to many future challenges.

Commercial and industrial

As firms increase their use of information technology and precision-controlled equipment, or just face a pared-down work environment with little slack in the schedule, poor power quality is becoming a central topic. The goal of commercial and industrial firms is to prevent their energy usage from overtly affecting their operations. Energy storage technologies currently assist by protecting against both power sags and outages in UPS products as a ride-through resource. Increasingly, a number of other energy storage technologies are following the lead of thermal energy storage technologies and providing a peak-shaving capability to these firms to reduce their energy costs as well as provide more control over their energy resources, ensuring that cost-saving measures like shifting activities to off-peak periods are kept to a minimum to reduce the impact on operations.

Utilities

Utilities have supported the development and deployment of energy storage technologies because these technologies promise to overcome problems with added peak capacity, system stability, transmission deferral, renewable integration, and customer power-quality issues. This continued interest has been supported by the flexible capabilities of these technologies to fit the changing needs of utilities. In the 1970s and 1980s, the market was dominated by large, vertically integrated utilities that needed a means to provide additional power resources during peak times cheaply—especially as a diurnal sink for the growing fleet of nuclear power plants. Although a number of technologies were evaluated (superconducting magnetic energy storage [SMES], compressed air energy storage [CAES], and pumped-hydroelectric storage [PHS]), pumped-hydro storage is becoming the most widely deployed. The 1990s brought the demand for more distributed facilities with a multifunctional capability to provide transmission-system stability and possibly asset deferral. Now utilities are increasingly evaluating other storage technologies that not only follow this multifunctional approach but increase their ability to be used as a working storage capacity (high cycling capability) to help provide stability in an unbalanced environment. Utilities are also envisioning storage technologies as a solution to the challenge of integrating power from wind farms as their penetration of their market increases.

Energy storage technologies hold great potential for their own uses as well as for those of other parties dealing with transmission system stability, such as ISOs and RTOs. Besides providing system stability for normal operations, they are assisting in the formation of ancillary services by creating much-needed price visibility. This came out of the experience with pumped-hydro facilities, where once they were installed to provide commodity energy arbitrage, they proved very capable of providing additional capabilities for the vertical utility. These other capabilities were soon highly prized by these utilities—including frequency regulation and contingency reserves. Now (and in the future), ISO and RTO development provide price discovery for ancillary services. This is important because in a vertical utility, all of the value was absorbed into the overall cost structure of the utility (that did not know the real value and costs of these services or its own marginal cost of service).

Finally, utilities have continued to view storage technologies as a solution to improving customer satisfaction. Beyond improving the quality of the power, these technologies can also help reduce the cost of delivering that power. For answers to both of these, utilities seek to install storage technologies either on their distribution system or at the customer site. An example of this is the continued support for thermal energy storage. The utilities can reduce energy costs for customers and help reduce peak demand, which allows them to prevent additional generation, transmission, and stability issues.

Energy service companies

Just as utilities are looking to storage technologies to improve the delivered power quality for customers, a growing numbers of energy service companies (ESCOs) are gaining interest in storage technologies to improve their customers' service quality. Many of these companies have three main goals for their efforts here. First, energy storage technologies can help ESCOs extend their service offering to their commercial and industrial end-use customers. Second, energy storage technologies provide for a demand for power electronics from a subsidiary or partner firm. Finally, energy storage technologies create additional demand for the firm's professional engineering services, whether it is site development in the initial stages or a continuing revenue stream through follow-on system maintenance contracts.

Regulators

Public-policy regulators also deem energy storage technologies to be tools for promoting their goals within the electric power industry, such as enabling a transition to a self-correcting market instead of one based on command and control regulation regime. By acting as a shock absorber, energy storage technologies provide optionality to all actors throughout the industry. In this way, energy storage technologies solve one of the regulators' greatest challenges—the need to find solutions for certain players that can be leveraged to other groups. For example, by helping utilities find an easier and inexpensive means to provide extended and enhanced transmission expansion, customers benefit through lower cost and higher service level, such as in the wholesale market by providing additional power resources during peak power. By supplying power throughout these stressed periods, even a small increase in resources can have a dramatic impact on power's spot price (fig. 6–2). The cumulative effect is that adding extra facilities spaced throughout the system actually provides an even greater capability to react to short-term power imbalances, which increases the effectiveness of large-scale energy storage facilities.

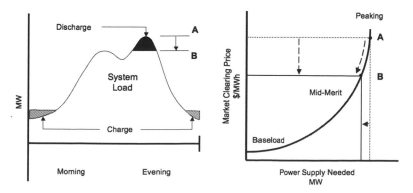

Fig. 6–2. Energy storage reduces peak prices (Ardour Capital Investments).

A Road Map for the Future

To continue the successful development of these energy storage technologies, three core issues must continue to be addressed to ensure the continued penetration of storage technologies into the power market. These include improving the technology, easing its deployment, and ensuring that the value in installing a storage unit continues to outweigh alternatives.

- **Technology.** Because of the number of different technologies included in energy storage technologies, the level of maturity ranges from just becoming commercially viable to mature with many years, even decades, of active use. However, all of these storage technologies will need to continue advancing their state of art, not only to improve their existing operating capabilities but also to provide the capacity to operate in new applications. Besides increasing the capability of the technology, increasing the level of reliability in operation is also vital.

- **Deployment.** Frequently customers are keenly interested in a particular technology, yet lack the ability to justify its initial costs or overcome other installation requirements. Government support and incentives have been crucial for driving the introduction of renewable energy and energy efficiency technologies; including energy storage technologies into these existing programs would yield significant results.

- **Value.** The first aspect of improving the value of any technology is to reduce the initial and operating costs of the technology, which have a large impact on life-cycle costs. These costs are reduced through design engineering, manufacturing processing, and the scale of production.

Other improvements, such as the ease of use, are important for the value others place on the technology—not just the consumer, but also system integrators contemplating incorporating a storage component into their overall system design. Finally, asset leveragability is important; when more components of the electric power industry are able to improve their own performance through coordination with a storage asset, it increases the value that asset has in the market.

Although the effort toward these areas of development can be driven primarily by the developer (in the case of technological development and some improvement in the ease of use), finding improvements in the deployment of the technology often will come through aligning the capabilities of storage technologies with the goals of other groups in the industry. This support can come from a variety of sources, including federal, state, and industry groups that see the capabilities of energy storage technologies furthering their causes, which will in turn help storage technologies overcome eventual hurdles.

Federal support

Learning the lessons of what storage can offer from other markets, federal policy makers are incorporating storage technologies into their plans to first move the market past its current challenges, and then to help maintain a self-correcting, competitive market once there. Because of the interaction of storage technologies with the electric power industry, this support will come in many forms as needed to assist with the development of the technology, providing a better market and reception for its deployment.

One has only to look as far as wind energy and energy efficiency technologies to see the use of federal support for a technology that has other attributes (in these cases, no emissions or fuel usage) that support goals valued by the federal government. This support (such as pilot projects to demonstrate the technology) has been crucial in developing alternatives to the existing, limited energy options. By providing support for commercialization of these and other advanced technologies, the federal government hopes to provide the market with a variety of tools to use their existing assets in new and cost-effective energy usage strategies to meet tomorrow's challenges.

The federal government has also had a significant role in deploying energy technologies into the market. Besides helping with accelerated depreciation of technology investment, or investment tax credits, the federal government has even assisted many technologies with export assistance. Other federal support, especially for the transmission market, includes recognizing the immense obstacles that exist for alternative technologies for existing T&D upgrades and promoting alternatives evenhandedly, so as not to skew investment decisions away from alternative strategies. Proponents of energy storage technologies suggest a number of steps to encourage the introduction of energy storage technologies into the transmission market:[4]

- Treat energy storage facilities the same as traditional transmission facilities for capacity expansion purposes (pricing incentives and planning processes).

- Have incentives to correct technical deficiencies separate from the regulatory process of forming RTOs.

- Prevent discrimination against or imposed disadvantages to alternatives to conventional transmission investments.

- Allow development costs for new technologies to be recovered in transmission rates to recognize the grid-wide benefits a strategically placed energy storage facility can provide.

- Award firm transmission rights and congestion revenue rights to parties that contract with energy storage service providers for services that substitute for or augment transmission capacity.

It is also possible to award these rights to parties that invest directly in energy storage systems that reduce congestion or otherwise produce transmission system benefits, such as permitting the deferral of transmission system investment.

- Do not prescribe the qualifying technology nor define narrowly the categories of improved performance that the Federal Energy Regulatory Commission (FERC) will reward.

State support

Although most attention points toward the federal government when the subject of governmental support for technology commercialization is mentioned, state governments have an important and increasing role to play. For technology development, for example, California (through the CEC) and New York (through NYSERDA) have supported near-commercial energy storage technologies by providing R&D grants. Because of the success of these programs in improving the quality of electric service to businesses and residential customers (plus encouraging high-tech energy technology firms to locate within the states), other state governments are looking to emulate California's and New York's successes with similar efforts of their own. Besides technology development assistance, state governments are also helping to deploy these technologies by supporting demonstration projects in real-world operating environments, and by lowering installation costs though incentives such as property tax reductions or other site cost advantages.

Even regulatory reform does not fall totally under the purview of the federal government. Each of the individual public utility commissions (PUCs) still controls much of the activity for serving customers with reliable and low-cost power. By influencing and supporting the introduction of energy storage facilities among utilities' distribution systems, PUCs potentially have even a greater influence on energy storage facility usage for system stability and asset deferral than the federal government.

One important means of supporting the introduction of energy storage technologies is to promote their use through state energy codes. In California, Title 24, Part 6 of the California Code of Regulations deals with Energy Efficiency Standards for Residential and Nonresidential Buildings. These regulations were established in 1978 in response to a legislative mandate to reduce California's energy consumption. The standards are updated periodically to incorporate new energy technologies and methods that promote these goals. The most recent changes, to take place October 1, 2005, will consider time-dependent valuation of energy, which in effect will favor peak energy saving measures over off-peak measures. These changes will give energy storage technologies that allow peak shaving significantly more support when being evaluated by a contractor during installation or refurbishment of a building. These building code changes have had a substantial impact on California's use of electricity. According to the state, California's building efficiency standards (along with those for energy efficient appliances) have saved more than $36 billion in electricity and natural gas costs since 1978. It is estimated the standards will save an additional $43 billion by 2013.[5]

Industry support

Besides governmental support and incentives, industry groups provide significant support and incentives to promote expansion of energy storage technologies in the market. Besides the support from EPRI for R&D and demonstration programs, other groups have been central to the development of energy storage technologies for some time. In particular, the Electricity Storage Association (ESA) has provided crucial technical and developmental support toward the commercialization of a wide variety of storage technologies for more than 14 years, and other groups such as the Energy Storage Council recently formed to advance public policy support for energy storage technologies.

To advance the presence of energy storage technologies throughout the market, however, energy storage technologies must be taken up by other industry groups and be seen as a means toward advancing their own goals. One successful example of industry support for energy storage technology lies in the commercial building market. The Leadership in Energy and Environmental Design (LEED) program is an excellent example of industry leaders recognizing the benefits of using thermal energy storage systems in their building designs. At the heart of LEED is the Green Building Rating System®—a voluntary, consensus-based national standard for developing high-performance, sustainable buildings. Through the LEED program, *green building* is defined as a common measurement standard to promote integrated, whole-building design practices. It provides a complete framework for assessing building performance and meeting sustainability goals, including energy efficiency. With thermal energy storage systems in the design, buildings may qualify for points in the "Energy and Atmosphere" section of the LEED's Green Building Rating System® toward the overall total score of the building. A higher score ensures not only lower overall operating costs of the building, but corresponds to a healthier building environment, and thus a greater demand for space in the building by potential tenants.

Many other potential areas exist for industry support for energy storage technologies; some of the most prominent being renewable energy proponents such as the American Wind Energy Association (AWEA). Providing support for renewable energy production has long been a goal of energy storage technology developers. Gaining the support of groups such as the AWEA would greatly accelerate the introduction of energy storage technologies to a host of potential installation opportunities. However, just as with the LEED program, energy storage technologies must continue to demonstrate that coupling an energy storage facility with a wind turbine accomplishes some goals better than what the turbine could achieve on its own. As explained in chapter 5, "Renewable Energy and Storage," there are many installations where wind turbines operate successfully without a storage facility. Storage developers should focus their initial attention on the technically or economically marginal installations where renewable energy resources can only be fully developed by adding energy storage facilities. If storage technology developers are able to demonstrate a means for wind developers to accomplish some goal that they were previously not able to accomplish without a storage component, then wind developers will be only too happy to include storage units in their planning, and will look for areas where they can be profitably incorporated into their plans.

Hurdles to overcome

Going forward, energy storage technologies as a class will need to overcome many obstacles to both continue developing their competitiveness and increase their market penetration. Some of the major hurdles include the following:

- Raising the visibility of the technology and its capabilities

- Proving the economic competitiveness of storage technologies

- Learning how to capture the value streams created by the facility

- Avoiding the trap of having the hype surrounding the technology overtake its realistic near-term capabilities and disappoint potential clients and adopters

Visibility. Gaining visibility (for its capabilities, its potential impacts, or even its existence) is important for any technology hoping to expand its presence in the market. One reason storage technologies have rarely garnered the spotlight so far is that most installations have operated behind the scenes in a system support, or standby role. Outside of the backup power roles of the ubiquitous battery, even the load-leveling roles of the pumped-hydro facility or commercial thermal storage units are not designed to operate as stand-alone sources of energy, but rather to extend the capabilities of other components by providing additional resources during the most capacity-constrained, highest-cost period of the day. Even when a storage unit is designed to act as a bridging power unit, it is only designed to maintain supply until a backup generator unit can be brought online.

Economics. Another hurdle energy storage technologies must overcome is justifying their investment economically. Proponents of energy storage technologies must prove an investment in these technologies is more valuable than allowing the current situation to continue, either through increasing revenue through active working storage or by reducing costs through T&D asset deferral or retail loss prevention. This proof must also be explained through conventional industry metrics used to evaluate any investment in technology, namely that it can provide a three-year (or less) payback and provide a reduction in life-cycle costs. Although up-front costs will always remain the main decision point for purchases, life-cycle costs are an important metric for firms able to take a variety of issues into consideration when purchasing technology. Finally, proponents of energy storage technologies need to show examples of follow-on value creation from the investment in energy storage technologies. This can include things like finding new or expanded uses for the technology once it has been installed. For example, many pumped-hydro facilities were built to provide commodity arbitrage, but once installed, an ancillary service market developed as a means to monetize additional capabilities such as frequency regulation, contingency reserves, and so forth. If the technology proves beneficial for additional parties, these groups may even subsidize the purchase. Here, thermal energy storage technologies were installed primarily to reduce a building's cooling load costs through peak shaving. However, this proved very beneficial to utilities in their attempt to reduce their load growth, so they now subsidize the investment in these storage technologies.

Multiple benefits to multiple beneficiaries. Another hurdle to overcome is that energy storage technologies' biggest strength is also their greatest challenge to widespread adoption—they provide multiple benefits to multiple beneficiaries. Most storage facilities (at all size levels) are multifunctional because they can absorb or discharge power at various rates. Because of the integrated nature of the power industry, all of the benefits of operating an energy storage facility do not fall directly to the owner. However, the final purchase or investment decision for any power industry technology is generally made on its capability in one function, not a basket of capabilities. Therefore, at least one capability must be more valuable than alternative applications—and enough so to warrant the purchase. Another aspect of this problem is that many of the applications of energy storage technologies are either not well-defined, or are easily captured by the owner of the facility. For example, in the wholesale power market, many of the benefits of a cache of stored power are because of the enhanced security of the network. Although some of this can be captured through the emerging ancillary service market, other benefits remain spread out among all market participants. For example, utilities looking to upgrade a power line only evaluate the capability of the energy storage facility to defer the upgrade versus that of capacity expansion—and they do not take into account the subsequent benefits toward system stability to later operations the energy storage facility brings.

Avoid being the "next big thing." Finally, although energy storage technologies hold out great potential for widespread capabilities throughout a number of invaluable applications, energy storage technologies could remain one of those perennial "next big things." As seen frequently in the past, new technologies like this are often heralded as the next "paradigm-busting disruptive technology" that will totally revolutionize the industry. Often, the most hard-core boosters for these new technologies envision the industry changing to its very foundation in only a matter of years. Unfortunately, the sheer size and complexity of the electric power industry are well-known. Even those nominally in charge of the system have scant control over how it truly operates—hence the push for massive investments for an intelligent grid with distributed reactive and proactive capabilities. Without such an understanding of its intended market, many of these paradigm-changing technologies fail to live up to their hype and become harder to site—even if it would be a beneficial and profitable placement.

Energy storage technologies have a real opportunity to avoid this trap and become integral components of the electric power value chain. Many next-big-thing technologies rely on being the center of activity for whatever system they become involved in. Energy storage technologies are different because they provide multiple benefits to multiple beneficiaries. They not only enable more than one activity to operate more efficiently and reliably, but they also provide widespread enhanced stability and security benefits. Finally, because these technologies promote an evolutionary rather than a revolutionary change to how the power industry operates, they are far more readily adoptable, and they make sense now—and in the future.

Right now, companies need a way to save money and free up capital for other desperately needed projects. Expanding the presence of energy storage systems now will help alleviate existing physical market constraints, improve overall operating economics, and provide relief to customers faced with increasingly unpredictable power quality and prices. In the future, these same firms will look for flexible tools to help them undertake new and creative business models that will emerge through competition. After the power market has transitioned to its future form (or continued to change as it has in the past), energy storage assets will still provide flexibility to market participants, freeing up resources for other purposes. By increasing the level of optionality in the market, storage will reinforce the market's self-regulating ability to reach policy makers' goals of a more efficient, reliable, and secure power industry. In some roles, large-scale energy storage technologies, acting as a shock absorber, will optimize existing generation facilities and stabilize power flows. In other uses, energy storage technologies prevent the catastrophic loss of vital information or disruption of an entire assembly line.

Energy storage technologies may not be the ultimate *disruptive* technologies with *paradigm-busting* results, but they are real technologies providing real results for real users.

References

1.   Collinson, A. 2000. *Electrical energy storage technologies for utility network optimization.* IEA Implementing Agreement on Energy Conservation through Energy Storage (Annex 9). Paris, France: International Energy Agency.

2.   *Distributed generation and electrical energy storage.* IEA Implementing Agreement on Energy Conservation through Energy Storage (Annex 15) (Proposed). Paris, France: International Energy Agency.

3.   *EPRI-DOE handbook of energy storage for transmission and distribution applications.* 2003. (EPRI Report # 1001834). Palo Alto, CA: Electric Power Research Institute.

4.   Bowe, J., and J. McLeod. Submitted March 13, 2003. *Comments of the Energy Storage Council on notice of proposed policy statement concerning establishment of incentives to promote efficient operation and expansion of the electric transmission grid* (FERC Docket No. PL03-1-000).

5.   California Energy Commission. Title 24, Part 6, of the California Code of Regulations: California's energy efficiency standards for residential and nonresidential buildings. http://www.energy.ca.gov/title24/ (accessed March 27, 2005).

Appendix

Vendor Web Sites

ABB	www.abb.com
Active Power	www.activepower.com
AFS Trinity	www.afstrinity.com
Alcad Limited	www.alcad.com
Alstom Power	www.power.alstom.com
American Hydro	www.ahydro.com
American Superconductor	www.amsuper.com
Baltimore AirCoil Company	www.baltimoreaircoil.com
Beacon Power	www.beaconpower.com
C&D Technologies, Inc.	www.cdtechno.com
CAES Development Company	www.caes.net
Calmac Manufacturing Corp.	www.calmac.com
Crown Battery Manufacturing Company	www.crownbattery.com
Cryogel	www.cryogel.com
Dresser-Rand	www.dr.com
Dunham-Bush, Inc.	www.dunham-bush.com
East Penn Manufacturing Company	www.dekabatteries.com
Evapco	www.evapco.com
Exide	www.exide.com

FAFCO, Inc.	www.fafco.com
Gill Batteries	www.gillbatteries.com
GNB Industrial Power	www.gnb.com
Hitec Power Protection	www.hitecups.com
Hoppecke Batterien GmbH	www.hoppecke.com
MWH	www.mwhglobal.com
NGK Insulators, Ltd.	www.ngk.co.jp
Paul Mueller Company	www.muel.com
Pentadyne Power Corporation	www.pentadyne.com
RWE Piller GmbH	www.piller.com
Plurion Systems, Inc.	www.plurionsystems.com
Premium Power Corporation	www.premiumpower.com
Saft	www.saft.com
SatCon Technologies Corporation	www.satcon.com
Sumitomo Electric Industries	www.sei.co.jp
Trojan Battery Company	www.trojanbattery.com
Tudor Batteries	www.tudorbatteries.com
Urenco Power Technologies	www.uptenergy.com
US Flywheel Systems	www.us-flywheel.com
VRB Power Systems	www.vrbpower.com
Vycon	www.calnetix.com
ZBB Energy	www.zbbenergy.com

Bibliography

Akhil, A., S. Swaminathan, and R. Sen. *Cost Analysis of Energy Storage for Electric Utility Applications*, Albuquerque, NM: Sandia National Laboratories, Feb.1997, Report SAND97-0443.

Akhil, A., and S. Kraft. *Battery Energy Storage Market Feasibility Study*, Albuquerque, NM: Sandia National Laboratories, Sept. 1997, Report SAND97-1275/1.

Alt, J., M. Anderson, and R. Jungst. *Assessment of Utility Side Cost Savings From Battery Energy Storage*, Albuquerque, NM: Sandia National Laboratories, 1995, Report SAND95-1545C.

Andersson, L., M. Nakhamkin, and R. Schainker. *AEC CAES Plant Delivers on Its Promises: Plant Performance and Lessons Learned*, presented at PowerGen International Conference, Dallas, Texas, Dec. 1997.

Annual Energy Outlook—2004, U.S. Department of Energy, Washington, DC, 2004.

Annual Energy Review—2002, U.S. Department of Energy, Washington, DC, 2004.

Baxter, R., and J. Makansi. *Energy Storage: The Sixth Dimension of the Electricity Value Chain*, Pearl Street, Inc., St. Louis, Missouri, Aug. 2002.

Baxter, Richard. "Electric Storage," *Electricity and Natural Gas Markets: Understand Them!* edited by Bob Willett, John Wiley & Sons, New York, 2003.

———. "Energy Storage: The Missing Link of the Electricity Value Chain," Energy Storage Council White Paper, St. Louis, MO, July 2002.

————. "Energy Storage in Today's Power Market," *World Power 2003*, Isherwood Production, London, UK, Mar. 2003.

————. "Energy Storage: Enabling Renewable Energy," *Renewable Energy World*, James & James, London, UK, July 2002.

Beckman, K., and P. Determeyer. "Natural Gas Storage: Historic Development and Expected Evolution," *GasTIPS*, Gas Research Institute (GRI), Washington, DC, June 1997.

Beno, J., J. Herbst, and R. Hebner. "Composite Flywheels for Energy Storage— Design Considerations," presentation at EESAT 2002, San Francisco, CA, Apr. 2002.

Blackaby, Nigel. "VRB Technology Comes to the Fore," *Power Engineering International*, PennWell Publishing, Tulsa, OK, May 2002.

Blankenship, Steve. "Technology Improving DG's Potential," *Power Engineering*, Jan. 2002.

Boyes, John. *Energy Storage Systems Program Report for FY99*, Albuquerque, NM: Sandia National Laboratories, 2000, Report SAND2000-1317.

Boyes, John. "Flywheel Energy Storage & Superconducting Magnetic Energy Storage Systems," presentation at the IEEE Power Engineering Society Summer Meeting, Seattle, WA, July 19, 2000.

————. "Overview of Energy Storage Applications," presentation at the IEEE Power Engineering Society Summer Meeting, Seattle, WA, July 19, 2000.

————. *Overview of Energy Storage Applications*, Albuquerque, NM: Sandia National Laboratories, June 2000, Report SAND2000-1040C.

————. *Technologies for Energy Storage Flywheels and Superconducting Magnetic Energy Storage*, Albuquerque, NM: Sandia National Laboratories, June 2000, Report SAND2000-1041C.

Bradshaw, D., and M. Ingram. "Pumped Hydroelectric Storage and Compressed Air Energy Storage," presentation at the IEEE Power Engineering Society Summer Meeting, Seattle, WA, July 19, 2000.

Bradshaw, D., M. Nakhamkin, R. Moritz, D. Hargreaves, R. Wolk, and M. DeLallo. "12 MW Cascaded Humidified Advanced Turbine (CHAT) Plant with 45–46% Efficiency Is Ready for Demonstration," presented at PowerGen International Conferencer, New Orleans, LA, December 1999.

Butler, P. *Battery Energy Storage for Utility Applications: Phase I—Opportunities Analysis*, Albuquerque, NM: Sandia National Laboratories, Nov. 1995, Report SAND95-2605.

Butler, P., J. Iannucci, and J. Eyer. *Innovative Business Cases for Energy Storage In a Restructured Electricity Marketplace: A Study for the DOE Energy Storage Systems Program*, Albuquerque, NM: Sandia National Laboratories, 2003, Report SAND2003-0362.

Butler, P., J. Miller, and P. Taylor. *Energy Storage Opportunities Analysis Phase II Final Report: A Study for the DOE Energy Storage Systems Program*, Albuquerque, NM: Sandia National Laboratories, 2002, Report SAND2002-1314.

Callahan, T., J. Degnan, and D. Miller. "Upgrading the Taum Sauk Pumped Storage Project," presentation at the Waterpower XIII meeting, Buffalo, New York: HCI Publications, 2003.

Cavallo, Alfred J. "Energy Storage Technologies for Utility Scale Intermittent Renewable Energy Systems," *Journal of Solar Energy Engineering*, 123, no. 4, Nov. 2001.

———. "High-capacity factor wind energy systems." *Journal of Solar Energy Engineering*, Transactions of the ASME, 117: 137–143, 1995.

Cavallo, A.J., R.H. Williams, and G. Terzian. "Baseload Wind Power from the Great Plains for Major Electricity Demand Centres." Center for Energy and Environmental Studies Report, Princeton University, Princeton, NJ, 25 pp. 1994.

Cavallo, A. and M.B. Keck. "Cost-Effective Seasonal Storage of Wind Energy," *Wind Energy*, SED-vol. 16 [Musial, W.D., S.M. Hock, and D.E. Berg (eds.)]. Book No. H00926-1995, American Society of Mechanical Engineers, 1995.

Clarke, Steve. "Observations on Building a Flow Battery Company," presentation at the Electricity Storage Association Conference, 2004.

———. "Introducing Cerium Based High Energy Redox Batteries," presentation at EESAT 2003 Conference, San Francisco, CA, 2003.

Commercial Potential of Natural Gas Storage in Lined Rock Caverns (LRC), Federal Energy Technology Center, U.S. Department of Energy, Washington, DC, Nov. 1999.

Corey, G., L. Stoddard, and R. Kerschen. *Boulder City Energy Storage Feasibility Study*, Albuquerque, NM: Sandia National Laboratory, 2002, Report SAND2002-0751.

Cowart, R. "Efficient Reliability: The Critical Role of Demand-Side Demand Resources in Power Systems and Markets," The National Association of Regulatory Utility Commissioners (NARUC), June 2001.

Crotogino, F., K. Mohmeyer, and R. Scharf. "Huntorf CAES: More than 20 Years of Successful Operation,"presentation at ASME Spring 2001 Meeting, Orlando, FL, Apr. 2001.

Daley, J., R. Loughlin, M. DeCorso, D. Moen, and L. Davis. "CAES—Reduced to Practice," presentation at the ASME Turbo EXPO 2001, June 2001.

DeAnda, M., and J. Miller. "Reliability of Valve-Regulated Lead Acid Batteries for Stationary Applications," presentation at EESAT 2002, San Francisco, CA, Apr. 2002.

DeAnda, M., J. Miller, P. Moseley, and P. Butler. *Reliability of Valve-Regulated Lead Acid Batteries for Stationary Applications*, Albuquerque, NM: Sandia National Laboratories, 2004, Report SAND2004-0914.

DeAnda, M., John D. Boyes, and W. Torres. *Lessons Learned from the Puerto Rico Battery Energy Storage System*, Albuquerque, NM: Sandia National Laboratories, Sept. 1999, Report SAND99-2232.

Deb, R., L. Hsue, A. Ornatsky, and J. Christian. *Surviving and Thriving in the RTO Revolution*, 2001. http://www.energyonline.com/reports/RTO_Revolution.pdf

Deb, Rajat K. "Operating Hydroelectric and Pumped Storage Units in a Competitive Environment," *The Electricity Journal*, April 2000, p 24–32.

Deb, R., P. Wagle, and R. Macatangay. "Generation and Transmission Investment in Restructured Electricity Markets," *The Environmental Monitor*, 2002. http://www.energyonline.com/reports/GTinvestments.pdf

Devries, Tim. "System Justification and Vendor Selection for the Golden Valley BESS," presentation at EESAT 2002, San Francisco, CA, Apr. 2002.

Dietert, J. A., and D. A. Pursell. *Underground Natural Gas Storage*, Simmons & Company International, Houston, TX, June 28, 2000.

Distributed Generation, edited by Borbely, Anne-Marie, CRC Press, New York, 2001.

Drouilhet, Steve M. "Energy Storage for Hybrid Village Power Systems," presentation at the Village Power '98 Technical Workshop, 1998.

———. "Power Flow Management in a High Penetration Wind-Diesel Hybrid Power System with Short-Term Energy Storage," presentation at Windpower '99, Burlington, VT, June 20–23, 1999.

Eisenstat, L., and M. Perlis. *Price Volatility in Wholesale Electricity Markets: Lessons Learned from the Midwest Price Spikes*, Electric Power Supply Association (EPSA), Washington, DC, Sept. 1998.

Electric Power Annual 2002, Volume 1 & 2, U.S. Department of Energy, Washington, DC, Nov. 2003.

Electricity Storage Association. Web site, http://www.energystorage.org.

Electricity Technology Roadmap: 2003 Summary and Synthesis, Electric Power Research Institute (EPRI), Palo Alto, CA, 2003.

Energy Infrastructure—Electricity Transmission Lines, Edison Electric Institute, Washington, DC, Feb. 2002.

Energy Infrastructure—Getting Electricity Where It Is Needed, Edison Electric Institute, Washington, DC, June 2001.

EPRI-DOE Handbook of Energy Storage for Transmission and Distribution Application, Electric Power Research Institute (EPRI) 1001834, Palo Alto, CA, Dec. 2003.

Eynon, R., T. Leckey, and R. Douglas Hale. *The Electric Transmission Network: A Multi-Regional Analysis*, U.S. Department of Energy, Washington, DC, 2001.

Fancher, R.B., *Dynamic Operating Benefits of Energy Storage*, Decision Focus, Inc.: Oct. 1986. Report EPRI AP-4875.

Fuldner, Arthur H. "Upgrading Transmission Capacity for Wholesale Electric Power Trade," U.S. Department of Energy, Washington, DC, 1999.

Gale, R., J. Graves, and J. Clapp. *The Future of Electric Transmission in the United States*, PA Consulting Group, Jan. 2001.

Gandy, Simon. "A Guide to the Range and Suitability of Electrical Energy Systems for Various Applications, and an Assessment of Possible Policy Effects," Imperial College of Science, Technology, and Medicine (University of London), Sept. 2000.

Gordon, S.P., and P. K. Falcone. *The Emerging Roles of Energy Storage in a Competitive Power Market: Summary of a DOE Workshop*, Albuquerque, NM: Sandia National Laboratories, June 1995, Report SAND95-8247.

"Grid 2030—A National Vision for Electricity's Second 100 Years," U.S. Department of Energy, Washington, DC, July 2003.

Grimsrud, P., S. Lefton, and P. Besuner. "Energy Storage Systems (ESS) Provide Significant Added Value by reducing the Cycling Costs of Conventional Generation," Aptech Engineering, Sunnyvale, CA, June 2004.

Guey-Lee, Louise. "Wind Energy Developments: Incentives in Selected Countries," Renewable Energy Annual 1998, U.S. Department of Energy, Washington, DC, Sunnyvale, CA, 2000.

Gyuk, Imre. "Electrical Energy Storage," presentation at the Energy Storage Association Meeting, Aug. 2000.

Handbook of Batteries, 3rd Edition, edited by D. Linden and T. Reddy, McGraw-Hill, NY, 2002.

Hassenzahl, W.V. *Energy Storage in a Restructured Electric Utility Industry— Report on EPRI Think Tanks I and II*: Electric Power Research Institute, Palo Alto, CA, Sept. 1997. Report EPRI TR-108894.

Hennessey, Timothy D.J. "The Multiple Benefits of Integrating the VRB-ESS with Wind Energy Producers—A Case Study in MWH Applications," presentation at the AWEA Conference 2004.

Hirst, Eric. *Bulk Power Ancillary Services for Industrial Customers*, Electricity Consumers Resource Council, June 1999.

———. *Expanding Transmission Capacity*, July 2000.

———. *Interaction of Wind Farms with Bulk-Power Operations and Markets*, Project for Sustainable FERC Energy Policy, Sept. 2001.

Hirst, E., and B. Kirby. "Ancillary Services: The Forgotten Issue," printed in *Electric Perspectives*, EEI, Washington, DC, July–Aug. 1998.

———. *Electric Power Ancillary Services*, Oak Ridge National Laboratory, Oak Ridge, TN, Feb.1996, Report ORNL/CON-426.

———. *Bulk Power Basics*, Mar. 12, 2000.

———. "Unbundling Generation and Transmission Services for Competitive Electricity Markets," Oak Ridge National Laboratory, Oak Ridge National Laboratory, Oak Ridge, TN, 1998, Report ORNL/CON-454.

———. "The Functions, Metrics, Costs, and Prices for Three Ancillary Services," Edison Electric Institute, Washington, DC, Oct. 1998.

———. "Measuring Generator Performance, in Providing the Regulation and Load-Following Ancillary Services," Dec. 2000.

Hopper, John M. "Can Natural Gas Storage Capacity Affect the Price of Electric Power? You'd Better Believe It!" *Power & Gas Marketing*, Oildom Publishing Company of Texas: Houston, TX, Jan./Feb. 2002.

Hrehor, R. and D. Sytsma. "Gas-Power Infrastructure: The Missing Link?" *Public Utilities Fortnightly*, Vienna, VA, Feb.15, 2001: 32–37.

Iannucci, J., B. Erdman, and J. Eyer. "Technical and Market Aspects of Innovative Storage Opportunities," presentation at EESAT 2002, San Francisco, CA, Apr. 2002.

Iannucci, J., and S. Schoenung. *Energy Storage Concepts for a Restructured Electric Utility Industry*, July 2000, Report SAND2000-1550.

Isherwood, W., R. Smith, S. Aceves, G. Berry, W. Clark, R. Johnson, D. Das, D. Goering, and R. Seifert. *Remote Power Systems with Advanced Storage Technologies for Remote Alaskan Villages*, National Renewable Energy Laboratory, Golden, CO, Dec. 1997, Report UCRL-ID-129289.

Jess, Margaret. "Restructuring Energy Industries: Lessons from Natural Gas," *Natural Gas Monthly*, U.S. Department of Energy, Washington, DC, May 1997, vii–xxi.

Kamibayashi, Makoto. "High Charge and Discharge Cycle Durability of the Sodium Sulfur Battery," presentation at EESAT 2002, San Francisco, CA, Apr. 2002.

Kroon, H., and G. H. C. M. Thijssen. "Determination of Commercial Viability of Flow Batteries," presentation at EESAT 2002, San Francisco, CA, Apr. 2002.

Lazarewicz, Matt. "A Description of the Beacon Power High Energy Power Composite Flywheel Energy Storage Systems," presentation at EESAT 2002, San Francisco, CA, Apr. 2002.

Lehr, Ronald C. *The National Wind Coordinating Committee Status Report and Regulatory Recommendations*, National Association of Regulatory Utility Commissioners, Aug. 2001.

Lotspeich, Chris. "A Comparative Assessment of Flow Battery Technologies," presentation at EESAT 2002 Conference, San Francisco, CA, Apr. 2002.

Maass, Peter. "Operation Experience with Huntorf, 290 MW World's First Air Storage System Energy Transfer (Asset) Plant," presentation at the American Power Conference, Apr. 21–23, 1980.

MacCracken, Mark. "Thermal Energy Storage Myths," *ASHRAE Journal* 45, no 9, Atlanta, GA, Sept., 2003: 36.

McDowall, Jim. "Battery Life Considerations in the Energy Storage Applications and their Effect on Life Cycle Costing," presentation at the IEEE Power Engineering Society, Summer Meeting, Vancouver, BC, 2001.

Makansi, Jason. "Energy Storage for Grid Management and RTO Management," presentation at Transmission: Operating and Planning Conference, Washington, DC, 2002.

Mears, Dan. *Introduction to NGK's Sodium Sulfur Batteries*, Technology Insights, July 2004.

Miller, R., and J. Malinowski. *Power System Operation*, 3rd edition, McGraw-Hill, New York, 1994.

Mitlitsky, F., B. Meyers, and A. H. Weisberg. *Regenerative Fuel Cell Systems R & D*, June 1998, Report UCRL-JC-131087.

Miyaki, Shinichi. "Vanadium Redox-Flow Batteries for a Variety of a Applications," presentation at the IEEE Power Engineering Society, Summer Meeting, Vancouver, BC, 2001.

Nakhamkin, M., and R. Wolk. "Compressed Air Inflates Gas Turbine Output," *Power Engineering*, PennWell Publishing, Tulsa, OK, Mar. 1999.

Nakhamkin, M., and R. Pot. *Hybrid Plant Technology for Distributed Power Generation*, presented at PowerGen International conference, New Orleans, LA, Dec. 1999.

National Transmission Grid Study, U.S. Department of Energy, Washington, DC, May 2002.

Natural Gas Annual 2002, U.S. Department of Energy, Washington, DC, Jan. 2004.

Natural Gas Infrastructure & Storage, Office of Fossil Energy, U.S. Department of Energy, Web site: http://www.fe.doe.gov/oil_gas/gasstorage/.

Natural Gas Storage—End User Interaction Task 2, Topical Report, Office of Fossil Energy, U.S. Department of Energy, Jan. 1996.

Nichols, David. "Electricity: Challenges and Solutions," presentation at the IEEE Power Engineering Society Winter Meeting, Columbus, OH, Jan. 29, 2001.

Nichols, D., B. Tamyurek, and H. Vollkommer. "Sodium Sulfur Battery (NAS) Applications," presentation at the IEEE Power Engineering Society Meeting, 2003.

Nourai, Ali. "Bulk Energy Storage Applications," presentation at the Electricity Supply Association Mini-Meeting, Washington, DC, Nov. 14, 2001.

Overview of Wind Technologies, U.S. Department of Energy, Washington, DC, 1997.

Pansini, A., and K. Smalling. *Basics of Electric Power Transmission*, PennWell Publishing Company, Tulsa, OK, 1998.

Parsons, B., M. Milligan, B. Zavadil, D. Brooks, B. Kirby, K. Dragoon, and J. Caldwell. *Grid Impacts of Wind Power: A Summary of Recent Studies in the United States*, National Renewable Energy Laboratory, Golden, CO, June 2003, Report NREL/CP-500-34318.

Patton, A.D., C. Singh, and D. Robinson. *The Impact of Restructuring Policy Changes on Power Grid Reliability*, Albuquerque, NM: Sandia National Laboratories, 1998. Report SAND98-2178.

Petersik, Thomas W. "Modeling the Cost of U.S. Wind Supply, Issues in Midterm Analysis for Modeling the Costs of U.S. Wind Supply," Washington, DC, 1999, Report EIA/DOE-0607(99).

Platt, C., and J. Hurwitch. "Energy Storage: It's not Just Load Leveling Anymore," *Public Utilities Fortnightly*, Vienna, VA, 136: 15, Aug. 1998, 52–56.

Price, A., G. Thijssen, and P. Symons. "Electricity Storage, A Solution in Network Operation?" presentation at DistribuTech Europe, Oct. 12, 2000.

Price, Anthony. "The Vulnerability of the Electricity Supply System," presentation at the Electricity Supply Association Mini-Meeting, Washington, DC, Nov. 14, 2001.

———. "Flow Batteries," presentation at the IEEE Power Engineering Society Summer Meeting, Seattle, WA, July 19, 2000.

Randazzo, Sam. "Grid Business: The Midwest Region," presentation at the Grid Business Conference, Mar. 2002.

Reliability Assessment 2002–2011, North American Electric Reliability Council, Princeton, NJ, Oct. 2002.

"Renewable Electricity Purchases: History and Recent Developments," *Renewable Energy Annual 1998 Issues and Trends*, U.S. Department of Energy, Washington, DC, 1998.

Renewable Energy Annual—2002, U.S. Department of Energy, Washington, DC, Nov. 2003.

Roberts, Brad. "Evolving Energy Storage Technologies for Protecting Large Scale Critical Loads," presentation at the IEEE Power Engineering Society Summer Meeting, Seattle, WA, July 19, 2000.

———. "High Power, Short Duration Energy Storage Systems," presentation at the IEEE Power Engineering Society Summer Meeting, Vancouver, BC, 2001.

Scaini, V., P. Lex, T. Rhea, and N. Clark. "Battery Energy Storage for Grid Support Applications," presentation at EESAT 2002, San Francisco, CA, Apr. 2002.

Schimmoller, Brian, "Compressed Air Energy Storage Makes Use of Space in Underground Caverns," *Power Engineering*, PennWell Publishing, Tulsa, OK, Aug. 1, 2001.

Schoenung, Susan. *Characteristics and Technologies for Long- vs. Short-term Energy Storage*, Albuquerque, NM: Sandia National Laboratories, Mar. 2001, Report SAND2001-0765.

Schoenung, S., and W. Hassenzahl. *Long- vs. Short-Term Energy Storage Technologies Analysis: A Life Cycle Cost Study*, Albuquerque, NM: Sandia National Laboratories, 2003, Report SAND2003-2783.

Sedano, Richard P. *Dimension of Reliability: A Paper on Electric System Reliability for Elected Officials*, National Council on Competition and the Electric Industry, Oct. 2001.

Shepard, S., and S. Van der Linden. "Compressed Air Energy Storage Adapts Proven Technology to Address Market Opportunities," *Power Engineering*, Apr. 2002, p. 34–37.

Shirazi, M., and S. Drouilhet. "An Analysis of the Performance Benefits of Short-Term Energy Storage in Wind-Diesel Hybrid Power Systems," presentation at the 1997 ASME Wind Energy Symposium, 1997, NREL/CP-440-22108.

Silvetti, B., and M. MacCracken. "Thermal Storage and Deregulation," *ASHRAE Journal*, Atlanta, GA, Apr. 1998.

Speaks, Kelly. "The Emerging Need for Intermediate Peaking Power," *Power and Gas Marketing*, Oildom Publishing Company of Texas, Houston Texas, Mar./Apr. 2002, 32–33.

Stack, James. "Development of a Performance Data Reduction, Analysis, and Reporting System for a Wind-Diesel Hybrid Power System," Office of Science, Department of Energy ERULF Program, Bucknell University/ National Renewable Energy Laboratory, Golden, CO, Aug. 2000.

"Study of Electric Transmission in Conjunction with Energy Storage Technology," Lower Colorado River Authority, Austin, TX, Aug. 2003.

Swaminathan, Shiva. *Review of Power Quality Applications of Energy Storage Systems*, Albuquerque, NM: Sandia National Laboratories, July 1998, Report SAND98-1513.

Swaminathan, S., N. Miller, and R. Sen. *Battery Energy Storage Systems Life Cycle Costs Case Studies*, Albuquerque, NM: Sandia National Laboratories, Aug. 1998, Report SAND98-1905.

Swaminathan, S., W. Flynn, and R. Sen. *Modeling of Battery Energy Storage in the National Energy Modeling System*, Albuquerque, NM: Sandia National Laboratories, Dec. 1997, Report SAND97-2926.

Symons, Philip C. "Opportunities for Energy Storage in Stressed Electricity Supply Systems," presentation at the IEEE Power Engineering Society, Summer Meeting, Vancouver, BC, 2001.

Taylor, P., L. Johnson, K. Reichart, P. DiPietro, J. Philip, and P. Butler. "A Summary of the State of the Art of Superconducting Magnetic Energy Storage Systems, Flywheel Energy Storage Systems, and Compressed Air Energy Storage System," July 1999, Albuquerque, NM: Sandia National Laboratories, Report SAND99-1854.

Taylor, R., D. Bradshaw, and J. Hoagland. "Energy Storage for Ancillary Services," presentation at EESAT 2002, San Francisco, CA, Apr. 2002.

Taylor, R., and J. Hoagland. "Using Energy Storage with Energy for Arbitrage," presentation at EESAT 2002, San Francisco, CA, Apr. 2002.

Thermal Energy Storage—Economics and Benefits, E3 Energy Services, LLC., Arlington, VA, 2002.

Thijssen, Gerard. "Supply Quality, Reliability & Storage," presentation at the Commercial Viable Electric Storage, London, UK, Jan. 30, 2001.

Torre, W., and V. Eckroad. "Improving Power Delivery Through the Application of Superconducting Magnetic Energy Storage," presentation at the IEEE Power Engineering Society Winter Meeting, Columbus, OH, Jan. 29, 2001.

"Transmission Pricing Issues for Electric Generation from Renewable Resources," *Renewable Energy Annual 1998 Issues and Trends*, U.S. Department of Energy, Washington, DC, 1998.

Transmission Reliability Multi-Year Program Plan FY 2001–2005, Office of Power Technologies, Energy Efficiency and Renewable Energy, U.S. Department of Energy, Washington, DC, May 2001.

The Value of Underground Storage in Today's Natural Gas Industry, U.S. Department of Energy, Washington, DC, Mar. 1995.

Van der Linden, Septimus. "CAES for Today's Market," presentation at EESAT 2002, San Francisco, CA, Apr. 2002.

———. "Compressed Air Energy Storage Adapts Proven Technology to Address Market Opportunities," *Power Engineering*, PennWell Publishing, Tulsa, OK, Apr. 2001.

Williams, Brad. "A Case Study: The Use of VRB Energy Storage System (VRB-ESS) as a Utility Network Planning Alternative," 14th Annual Electricity Storage Association Conference, 2004.

Wilson, Roger. "Market Driven CAES," presentation at Electric Power 2002, Mar. 2002.

Wind, Thomas. "Compressed Air Energy Storage in Iowa," presentation for the Iowa Association of Municipal Utilities, Wind Energy Consulting, 2002.

Wolsky, A.M. "Introduction to Progress and Promise of Superconductivity for Energy Storage in the Electric Power Sector," May 1998, Report ANL/ES/CP–96555.

Workshop on Electric Transmission Reliability, Workshop Report, U.S. Department of Energy Transmission Reliability Program and the Consortium for Electric Reliability Technology Solutions, Sept. 1999.

World UPS Markets: Alternative Energy Storage Solutions, San Jose, CA, Frost & Sullivan, 2003.

World Lead Acid Battery Markets, San Jose, CA, Frost & Sullivan, 2002.

Zink, J.C. "Who Says You Can't Store Electricity?" *Power Engineering*, PennWell Publishing, Tulsa, OK, Mar. 1997, 21–25.

Zucchet, Michael J. "Renewable Resource Electricity in the Changing Regulatory Environment," *Renewable Energy Annual 1995*, U.S. Department of Energy, Washington, DC, 1996.

Index

About the Author

RICHARD BAXTER is a Senior Technology Analyst with Ardour Capital Investments, an investment bank specializing in energy technologies and alternative energy markets. In this role, Richard specializes in evaluating energy technologies and their related business models. He is the author of numerous reports, industry journal articles, and is an accomplished public speaker. Before joining Ardour, Richard was the Director of Member Affairs at the Energy Storage Council where he was involved in raising the visibility of energy storage technologies for federal and state government officials. Prior to this, Richard led the electric power research at the Yankee Group, evaluating growth prospects for new technology firms and providing due diligence services to the financial industry. Richard also spent time at the Standard & Poor's DRI Energy Group where he was responsible for key environmental and electric power components of the U.S. electric power forecasts for senior utility executives on emerging technical and market changes. Richard has a M.S. degree in Energy Management and Policy from the University of Pennsylvania, a B.S. in History from Tennessee State University, and a B.S. in Materials Engineering from Virginia Tech.